Studies in Computational Intelligence

Volume 679

Series editor

Janusz Kacprzyk, Polish Academy of Sciences, Warsaw, Poland
e-mail: kacprzyk@ibspan.waw.pl

About this Series

The series "Studies in Computational Intelligence" (SCI) publishes new developments and advances in the various areas of computational intelligence—quickly and with a high quality. The intent is to cover the theory, applications, and design methods of computational intelligence, as embedded in the fields of engineering, computer science, physics and life sciences, as well as the methodologies behind them. The series contains monographs, lecture notes and edited volumes in computational intelligence spanning the areas of neural networks, connectionist systems, genetic algorithms, evolutionary computation, artificial intelligence, cellular automata, self-organizing systems, soft computing, fuzzy systems, and hybrid intelligent systems. Of particular value to both the contributors and the readership are the short publication timeframe and the worldwide distribution, which enable both wide and rapid dissemination of research output.

More information about this series at http://www.springer.com/series/7092

Oliver Kramer

Genetic Algorithm Essentials

 Springer

Oliver Kramer
Department for Computing Science,
    Computational Intelligence Group
University of Oldenburg
Oldenburg
Germany

ISSN 1860-949X   ISSN 1860-9503   (electronic)
Studies in Computational Intelligence
ISBN 978-3-319-84834-1   ISBN 978-3-319-52156-5   (eBook)
DOI 10.1007/978-3-319-52156-5

Printed on acid-free paper

This Springer imprint is published by Springer Nature
The registered company is Springer International Publishing AG
The registered company address is: Gewerbestrasse 11, 6330 Cham, Switzerland

Contents

Part II   Solution Spaces

Part III   Advanced Concepts

Abstract

GENETIC ALGORITHMS (GAs) are biologically inspired methods for optimization. In the last decades, they have grown to exceptionally successful means for solving optimization problems. *Genetic Algorithm Essentials* gives an introduction to GENETIC ALGORITHMS with an emphasis on an easy understanding of the main concepts, most important algorithms, and state-of-the-art applications. The depiction has three unique characteristics: It does not get lost in unnecessary details, it considers latest developments like machine learning for evolutionary search, and it abstains from an overload of formalisms and notations and thus opens the doors to a broader audience. The first part of this book gives an introduction to GENETIC ALGORITHMS starting with basic concepts like evolutionary operators. It continues with an overview of strategies for tuning and controlling parameters. The second part is dedicated to solution space variants such as multimodal, constrained, and multi-objective solution spaces. The third part gives a short introduction to theoretical tools for GENETIC ALGORITHMS, the intersections, and hybridizations with machine learning and shows a choice of interesting applications.

Part I
Foundations

Chapter 1
Introduction

1.1 Optimization

This book gives an introduction to concepts and ideas of GENETIC ALGORITHMS. Before it begins, it is reasonable to clarify, what kinds of problems are solved with GENETIC ALGORITHMS. The answer is simple and short: optimization problems. Optimization is the task of finding optimal solutions, which are solutions that have a better quality than others. We often seek for the global optimal solution, which is the best solution in the whole solution space. This can be a tedious task, as the solution space can suffer from constraints, noise, strange fitness function conditions, unsteadiness, and a large number of local optima. If modeled in an appropriate kind of way, GENETIC ALGORITHMS are able to solve most optimization problems that occur in practice.

Optimization problems can be found in many domains, from natural sciences to math and computer science, from engineering to social and daily life. Whenever the task is to minimize an error, to minimize energy, weight, waste, effort and to maximize profit, outcome, success, and scores, we face optimization problems.

There are many famous optimization problems in computer science with efficient algorithms that have been proposed to solve them. For many hard problems no efficient solution is available and heuristics like GENETIC ALGORITHMS are reasonable to apply. The traveling salesman problem is an example for a hard optimization problem, for which heuristics deliver an acceptable solution in practice. In short, the traveling salesman problem seeks for a permutation of cities, such that the length of the tour the salesman has to travel is the shortest. Every city is only allowed to be visited once except of the starting point that has to be reached at the end of the tour again. In other words, we seek for the shortest round trip between a set of cities while visiting each city only once. As the number of possible permutations for this round trip grows exponentially with the number of cities, this problem is difficult to solve.

In GENETIC ALGORITHM research artificial benchmark problems are used for experimental research. These functions are explicitly given and analytically solvable. Hence, their characteristics, their structure, and their optima are well known. Given

© Springer International Publishing AG 2017
O. Kramer, *Genetic Algorithm Essentials*, Studies in Computational
Intelligence 679, DOI 10.1007/978-3-319-52156-5_1

Fig. 1.1 Fitness landscape
of the Sphere function
$f(\mathbf{x}) = \sum_i x_i^2$

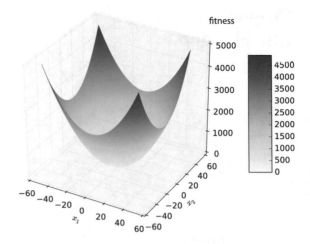

this information, we can test, if your algorithms are able to solve the corresponding problems. We can challenge the optimization algorithms. A famous and often applied example in continuous optimization is the Sphere function, see Fig. 1.1. It is simply structured and employs only one global optimum in the origin. A GENETIC ALGO-RITHM for continuous solution spaces should be able to approximate the optimum fast and with almost arbitrary accuracy.

GENETIC ALGORITHMS are the translation of the biological concept of evolution into algorithmic recipes. They belong to the area of computer science related to machines and computer programs. As they are part of many intelligent systems, GENETIC ALGORITHMS are frequently counted to the areas of computational intelligence and artificial intelligence, which aim at constructing methods that imitate and even overcome human intelligence. Meanwhile, a huge collection of methods has been proposed that fall into these categories.

Classic computational intelligence comprises the three branches GENETIC ALGO-RITHMS, neural networks [88], and fuzzy logic, which have in common to be nature-inspired. Meanwhile, artificial immune systems and swarm intelligence have also become important areas in computational intelligence. Artificial intelligence is a term more related to symbol-oriented algorithms that solve human problems like propositional logic, planning strategies, and shortest path algorithms. Machine learning concentrates on methods for learning from data, in particular methods that cover the problem classes classification, clustering, and dimensionality reduction. The same holds for data mining, which is machine learning with an emphasis on data bases and a very large set of data samples.

To summarize, GENETIC ALGORITHMS are excellent methods for hard optimization problems, where classic optimization methods fail due to difficult characteristics. Such conditions can be unsteadiness, non-derivability, noise and many other. In the course of this book it will become clear, why GENETIC ALGORITHMS are capable of handling such conditions.

1.2 From Biology to Genetic Algorithms

GENETIC ALGORITHMS are biologically-inspired algorithms for optimization. In his famous work *On the Origin of Species* Charles Darwin was the first, who proposed the concept of evolution [13]. It is an explanation for the biological development of species with mating selection and survival of the fittest. Evolution developed a representation known as deoxyribose nucleic acid (DNA). The DNA encodes creatures and is the basis for evolutionary processes. In other words, DNA is the representation for biological life. Creatures are the best example for showing that natural evolution is a successful optimization process that has been running since four billion years. The current development of species might also be subject to a longer optimization process, if genetic material has been carried to earth via asteroids.

The time resolution of evolution varies remarkably. New species can arise in weeks or even days like bacteria while the evolution of other species remains stable for long periods. In contrast, turtles are an example for slow and stable evolution. The fast evolution of bacteria can be explained with its low structural complexity and the fast reproduction rate. The genome encodes proteins that induce biological processes in cells and organisms.

Figure 1.2 visualizes the continuous cycle of artificial evolution that is based on the principles of natural evolution. The evolutionary process begins with randomly or manually initialized solutions. The evolutionary cycle starts by recombining two or more solutions with the crossover operator. The outcome is mutated. The best solutions that have been generated this way are selected for the following generation. Last, the evolutionary cycle examines, if the termination condition has been met, and continues the genetic optimization run, if this is not the case yet.

Usually, a population of solutions is employed. But the simplest variant of GENETIC ALGORITHMS is the (1+1)-GENETIC ALGORITHM that is only based on one parent that is mutated to a child. The selection operator chooses the better solution, which can be the parent or the child. Recombination is not applied as only one parent exists in each generation. For almost all kinds of solution representations crossover and mutation operators can be designed. In the course of this book some

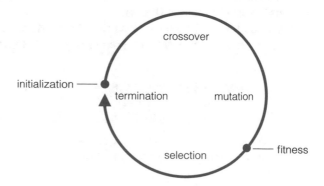

Fig. 1.2 GENETIC ALGORITHM cycle of initialization, crossover, mutation, fitness computation, selection, and termination

variants will be introduced like continuous vectors, bit strings, and permutations of symbols.

1.3   Genetic Algorithm Variants

The development of algorithms that are oriented to evolution began in the sixties of the 20th century. Four main streams of GENETIC ALGORITHM variants have developed almost independently. Nowadays, they count to the family of GENETIC ALGORITHMS, see Fig. 1.3. This is why the question for GENETIC ALGORITHM variants is deeply connected to their history. Ingo Rechenberg and Hans-Paul Schwefel evolved artificial systems with algorithms they called evolution strategies in Europe [86, 93]. This class of GENETIC ALGORITHMS is still a famous research branch with an emphasis on continuous solution spaces. Gaussian mutation [5] is combined with mutation rate adaptation mechanisms like Rechenberg's 1/5th rule [86] or self-adaptation [94]. The latter allows an automatic control of the mutation rates.

Also in the sixties of the 20th century John Holland introduced GENETIC ALGORITHMS as optimization methods in the United States [39]. The first GENETIC ALGORITHMS were mainly based on binary string representations. A decoder function is required for mapping the bit string genotype to the phenotype, which is finally the solution to the particular problem. Crossover operators played a more important role than mutation in the early days of Holland's GENETIC ALGORITHMS. Mutation was mainly bit flip mutation flipping zeros to ones and vice versa with a fixed probability.

Moreover, Fogel, Owens and Walsh introduced evolutionary programming [27], which was originally designed for evolving deterministic finite automata that accept a set of input strings. Later, evolutionary programming was extended for optimization in binary and continuous solution spaces as well, also equipped with mutation rate adaptation techniques.

Today all variants have grown together. It is hardly possible to distinguish different variants, because most concepts, representations, and mechanisms have been introduced to all algorithmic variants. But there are still tracks on the main GENETIC ALGORITHM conferences focusing on special solution space characteristics like continuous spaces in evolution strategies.

Only the fourth branch of GENETIC ALGORITHMS can still be distinguished from the other variants: genetic programming [3, 49]. Genetic programming evolves

Fig. 1.3 Overview of GENETIC ALGORITHM variants. GENETIC ALGORITHMS, evolution strategies, evolutionary programming, and genetic programming belong to the same family of optimization algorithms

machine learning programs. Hence, the main difference to other GENETIC ALGO-
RITHMS is the representation. Programs can be represented in many ways, for example
as trees or as assembler programs. During the evaluation of the solution quality, the
program is run and its performance is measured.

1.4   Related Optimization Heuristics

Numerous optimization heuristics have been proposed in the last decades that are similar to GENETIC ALGORITHMS. Most of them have in common that they are inspired
by natural processes and that they are based on stochastic operators. Furthermore,
all have in common that they employ exploration and exploitation mechanisms.
Simulated annealing is based on mutating solutions and accepting them in case of
improvements. With a certain probability based on a decreasing parameter called
temperature worse offspring is accepted [48].

Very closely related to GENETIC ALGORITHMS is swarm intelligence [7]. Swarm
algorithms are inspired by natural swarms like birds, bees, and ants and are consequently also based on populations of individuals. Particle swarm optimization
algorithms mimic the movement of flocks to move in solution space [47]. Oriented
to the movements of their neighbors and the best positions in solution space found so
far, particles move in solution space with velocities and randomness. This plays the
role of the explorative mechanism. The orientation to the best so far found positions
is the exploitative counterpart.

Ant colony optimization algorithms are search algorithms well appropriate for
combinatorial optimization problems [20]. Based on the idea of pheromones that
ants leave on the surfaces while finding the shortest way from a nest to food sources,
ant colony optimization algorithms distribute pheromones as reward signals on useful
solution fragments depending on the quality of the overall solution.

Fireworks algorithms imitate the movements and dynamic of fireworks [98]. A
shower of sparks fills the local space around the firework. This concepts is very
related to GENETIC ALGORITHMS as it tries to find optimal locations of fireworks
while iteratively generating sparks, similar to mutation, evaluating each location and
finally choosing the best location as basis for the following iteration. Recently, a
mechanism has been proposed for balancing exploration and exploitation [11].

Firefly algorithms mimic the flashing behavior of fireflies [106]. The idea is that
unisexual fireflies are attracted by the brightness of solutions that reflects their quality
decreasing with increasing distances. If no firefly is brighter than a defined distance
is exceeded, fireflies randomly move in space.

1.5   This Book

This book is an introduction to GENETIC ALGORITHMS. GENETIC ALGORITHMS are biologically inspired optimization heuristics. Meanwhile established as solid methodological playground, in particular for difficult optimization problems, a long line of research has proven their success, experimentally and theoretically. Consequently, the list of topics associated with GENETIC ALGORITHMS has grown, which is surely overwhelming for beginners in the field. But also the expert may sometimes be surprised about the variety of areas and evolving fields. Figure 1.4 visualizes a list of typical keywords that are associated with GENETIC ALGORITHMS and that play a role in the course of this book reaching from genetic operators like mutation and selection to applications like learning and constraint handling. The variety of topics is large and the list of research papers in each field is overwhelming. However, when having a closer look, many concepts are similar to each other, a baseline of similar mechanisms and heuristics exists. It is the task of this book to identify the most important ones and to compose them to a comprehensive depiction.

There are many good related depictions. Most of them put another focus on the introductory concepts. Eiben and Smith [23] give a good overview to GENETIC ALGORITHMS. The book is an attractive introduction of GENETIC ALGORITHM concepts and gives an overview to many related fields. A recommendation to all researchers, who are interested in theoretical results, is the book by Neumann and Witt [80]. It introduces basic proof techniques and exemplary shows interesting runtime analysis results. An introduction to the related field of machine learning is given by James et al. [42], which is an introductory variant of the famous counterpart *Elements of statistical learning* [38].

This book is structured as follows.

- Chapter 2 gives an introduction to GENETIC ALGORITHMS concentrating on the most important concepts like populations, the generational scheme, crossover, mutation, selection, genotype-phenotype mapping, and termination conditions.
- Chapter 3 puts an emphasis on tuning and control of GENETIC ALGORITHMS' parameters. It turns out that the genetic search crucially depends on the choice of adequate parameters. The chapter gives an introduction to different tuning and control strategies like dynamic control, Rechenberg's mutation rate control, and self-adaptation.

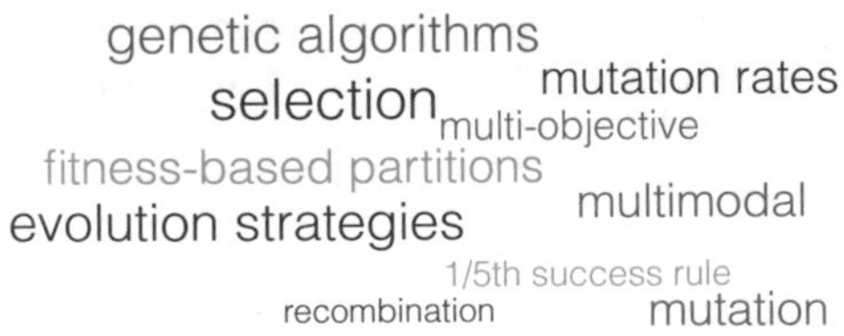

Fig. 1.4 Cloud of typical keywords associated with GENETIC ALGORITHMS

genetic algorithms
mutation rates
selection multi-objective
fitness-based partitions
evolution strategies multimodal
1/5th success rule
recombination mutation

- Chapter 4 present strategies for overcoming local optima in order to approximate the global one, or at least to find as many local optima as possible. Strategies comprise restarts, niching, fitness sharing, and novelty search.
- Chapter 5 gives an introduction to constraint handling. Practical optimization problems are often constrained. Not the whole solution space is allowed, but only a feasible subset. GENETIC ALGORITHMS have to be adapted to cope with constraints. One strategy is to choose representations and operators that avoid infeasible solutions. An easy way to cope with constraints is the use of penalty functions, which deteriorate the fitness of a solution.
- Chapter 6 presents multi-objective optimization approaches. Multiple optimization goals induce multiple objectives. If they are conflictive, the minimization of one objective results in the maximization of another. Strategies like the non-dominated sorting GENETIC ALGORITHM, rake selection, and selection based on the hypervolume indicator allow the approximation of a Pareto-set of solutions.
- Chapter 7 gives an introduction to theoretical research on GENETIC ALGORITHMS. Starting with an overview of theoretical research in this area, a runtime analysis is exemplarily presented and an overview of further theoretical tools is presented.
- Chapter 8 presents research in the intersection between machine learning and GENETIC ALGORITHMS. Machine learning algorithms can be used to improve and support genetic search. Examples include covariance matrix estimation, meta-modeling, and visualization. GENETIC ALGORITHMS are effective tuning and learning approaches for machine learning problems.
- Chapter 9 shows GENETIC ALGORITHMS in various applications. GENETIC ALGORITHMS can be used to optimize machine learning problems, to optimize wind turbine locations considering wake effects and geo-constraints, to scale the features of wind power data in nearest neighbor regression, and to optimize rule bases for virtual power plants.
- Finally, Chap. 10 closes with a summary of the most important aspects and contributions of this book.

The book closes with a summary and an appendix containing supplementary information.

1.6 Further Remarks

This depiction passes on complex notations where possible. Although a thorough mathematical formulation might be much more exact helping to implement and model certain aspects and details, it also can complicate the understanding of concepts, the context, and the connections between different mechanisms. However, here is some basis notation that we usually employ in our depictions. A vector is a bold small letter $\mathbf{x}$ while a scalar is written with a small plain letter.

Also the pseudocode of most algorithms presented in this book passes on complex notations. More important than the interpretation of possibly arbitrary levels of abstractions and formalisms is the understanding of the introduced concepts. This is provably easier for beginners, if the depiction is easy.

The figures throughout this book have been created with Apple Keynote. Plots have been generated with Python and matplotlib [41]. As Python allows fast prototyping of new ideas and concepts, it is the recommended programming language in our related lectures and research activities. Further, the employment of Python-based machine learning methods is recommended, for example for meta-modeling, covariance matrix estimation, and dimensionality reduction. A recommended machine learning library is sklearn [84], which is also extensively used in [56]. Sklearn contains implementations of most state-of-the-art machine learning methods like support vector machines, k-means clustering, and neural networks. The library is steadily improved and extended.

Chapter 2
Genetic Algorithms

2.1 Introduction

GENETIC ALGORITHMS are heuristic search approaches that are applicable to a wide range of optimization problems. This flexibility makes them attractive for many optimization problems in practice. Evolution is the basis of GENETIC ALGORITHMS. The current variety and success of species is a good reason for believing in the power of evolution. Species are able to adapt to their environment. They have developed to complex structures that allow the survival in different kinds of environments. Mating and getting offspring to evolve belong to the main principles of the success of evolution. These are good reasons for adapting evolutionary principles to solving optimization problems.

In this chapter we will introduce the foundations of GENETIC ALGORITHMS. Starting with an introduction to the basic GENETIC ALGORITHM with populations, we will introduce the most important genetic operators step by step, which are crossover, mutation, and selection. Further, we will discuss genotype-phenotype mapping, common termination conditions, and give a short excursus to experimental analysis.

2.2 Basic Genetic Algorithm

The classic GENETIC ALGORITHM is based on a set of candidate solutions that represent a solution to the optimization problem we want to solve. A solution is a potential candidate for an optimum of the optimization problem. Its representation plays an important role, as the representation determines the choice of the genetic operators. Representations are usually lists of values and are more generally based on sets of symbols. If they are continuous, they are called vectors, if they consist of bits, they are called bit strings. In case of combinatorial problems the solutions often consist of symbols that appear in a list. An example is the representation of a tour in case of the traveling salesman problem. Genetic operators produce new solutions in the chosen

© Springer International Publishing AG 2017
O. Kramer, *Genetic Algorithm Essentials*, Studies in Computational
Intelligence 679, DOI 10.1007/978-3-319-52156-5_2

representation and allow the walk in solution space. The coding of the solution as representation, which is subject to the evolutionary process, is called genotype or chromosome.

Algorithm 1 shows the pseudocode of the basic GENETIC ALGORITHM, which can serve as blueprint for many related approaches. At the beginning, a set of solutions, which is denoted as population, is initialized. This initialization is recommended to randomly cover the whole solution space or to model and incorporate expert knowledge. The representation determines the initialization process. For bit string representations a random combination of zeros and ones is reasonable, for example the initial random chromosome 1001001001 as a typical bit string of length 10. The main generational loop of the GENETIC ALGORITHM generates new offspring candidate solutions with crossover and mutation until the population is complete.

Algorithm 1 Basic GENETIC ALGORITHM

1: initialize population
2: **repeat**
3:     **repeat**
4:         crossover
5:         mutation
6:         phenotype mapping
7:         fitness computation
8:     **until** population complete
9:     selection of parental population
10: **until** termination condition

2.3 Crossover

Crossover is an operator that allows the combination of the genetic material of two or more solutions [97]. In nature most species have two parents. Some exceptions do not know different sexes and therefore only have one parent. In GENETIC ALGORITHMS we can even extend the crossover operators to more than two parents. The first step in nature is the selection of a potential mate partner. Many species spend a lot of resources on selection processes, but also on the choice of a potential partner and on strategies to attract partners. In particular, males spend many resources on impressing females. After the selection of a partner, pairing is the next natural step. From a biological perspective, two partners of the same species combine their genetic material and inherit it to their offspring.

Crossover operators in GENETIC ALGORITHMS implement a mechanism that mixes the genetic material of the parents. A famous one for bit string representation is n-point crossover. It splits up two solution at n positions and alternately assembles them to a new one (Fig. 2.1). For example, if 0010110010 is the first parent and 1111010111 is the second one, one-point crossover would randomly choose a position, let us assume 4, and generate the two offspring candidate solutions 0010-

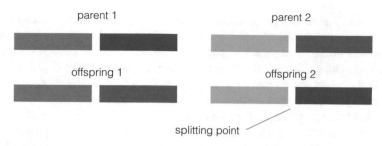

Fig. 2.1 Illustration of one-point crossover that splits up the genome of two solutions at an arbitrary point (here in the *middle*) and re-assembles them to get two novel solutions

010111 and 1111-110010. The motivation for such an operator is that both strings might represent successful parts of solutions that when combined even outperform their parents. This operator can easily be extended to more points, where the solutions are split up and reassembled alternately.

For continuous representations, the crossover operators are oriented to numerical operations. Arithmetic crossover, also known as intermediate crossover, computes the arithmetic mean of all parental solutions component-wise. For example, for the two parents $(1, 4, 2)$ and $(3, 2, 3)$ the offspring solution is $(2, 3, 2.5)$. This crossover operator can be extended to more than two parents. Dominant crossover successively chooses each component from one of the parental solutions. Uniform crossover by Syswerda [97] uses a fix mixing ratio like 0.5 to randomly choose a bit from either of the parents. The question comes up, which of the parental solutions take part in the generation of new solutions. Many GENETIC ALGORITHMS simplify this step and randomly choose the parents for the crossover operation with uniform distribution.

2.4 Mutation

The second protagonist in GENETIC ALGORITHMS is mutation. Mutation operators change a solution by disturbing them. Mutation is based on random changes. The strength of this disturbance is called mutation rate. In continuous solution spaces the mutation rate is also known as step size.

There are three main requirements for mutation operators. The first condition is reachability. Each point in solution space must be reachable from an arbitrary point in solution space. An example that may complicate the fulfillment of this condition is the existence of constraints that shrink the whole solution space to a feasible subset. There must be a minimum chance to reach every part of the solution space. Otherwise, the chance is not positive that the optimum can be found. Not every mutation operator can guarantee this condition, for example decoder approaches have difficulties covering the whole solution space.

The second good design principle of mutation operators is unbiasedness. The mutation operator should not induce a drift of the search to a particular direction, at least in unconstrained solution spaces without plateaus. In case of constrained solution spaces bias can be advantageous, which has been shown in [50, 62]. Also the idea of novelty search that tries to search in parts of the solution space that are unexplored yet, see Chap. 4, induces a bias on the mutation operator.

The third design principle for mutation operators is scalability. Each mutation operator should offer the degree of freedom that its strength is adaptable. This is usually possible for mutation operators that are based on a probability distribution. For example, for the Gaussian mutation that is based on the Gaussian distribution the standard deviation can scale the randomly drawn samples in the whole solution space. The implementation of the mutation operators depends on the employed representation. For bit strings bit flip mutation is usually used. Bit flip mutation flips a zero bit to a one bit and vice versa with a defined probability, which plays the role of the mutation rate. It is usually chosen according to the length of the representation. If N is the length of the bit string, each bit is flipped with mutation rate $1/N$. In Chap. 7 we will present a runtime analysis that is based on bit flip mutation. If the representation is a list or string of arbitrary elements, mutation randomly chooses a replacement for each element. This mutation operator is known as random resetting. Let $[5, 7, -3, 2]$ be the chromosome with integer values that come from the interval $[-10, 10]$, then random resetting decides for each component, if it is replaced. If the component is replaced, it randomly chooses a new value from the interval. For example, the result can be $[8, -2, -5, 6]$.

For continuous representations, Gaussian mutation is the most popular operator. Most processes in nature follow a Gaussian distribution, see Fig. 2.2. This is a reasonable assumption for the distribution of successful solutions.

A vector of Gaussian noise is added to a continuous solution vector [5]. If $\mathbf{x}$ is the offspring solution that has been generated with crossover,

$$\mathbf{x}' = \mathbf{x} + \sigma \cdot \mathcal{N}(0, 1) \tag{2.1}$$

Fig. 2.2 The Gaussian distribution is basis of the Gaussian mutation operator adding noise to each component of the chromosome

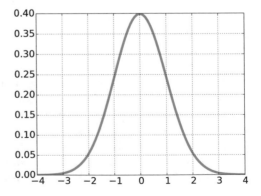

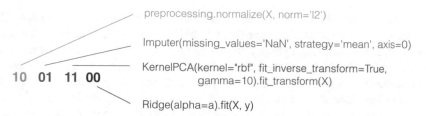

Fig. 2.3 Example of genotype-phenotype mapping for a machine learning pipeline. The bit string encodes a pipeline of normalization, imputation, dimensionality reduction, and regression

is the Gaussian mutation with $\mathcal{N}(0, 1)$ as notation for a vector of Gaussian-based noise. Variable σ is the mutation rate that scales the strengths of the noise added. The Gaussian distribution is maximal at the origin. Hence, with the highest probability the solution is not changed or only slightly. The Gaussian mutation is an excellent example for a mutation operator that fulfills all mentioned conditions. With σ it is arbitrarily scalable. Moreover, with a scalable σ, all regions in continuous solution spaces will be reachable. Due to the symmetry of the Gaussian distribution, it does not prefer any direction and is hence driftless.

2.5 Genotype-Phenotype Mapping

After crossover and mutation, the new offspring population has to be evaluated. Each candidate solution has to be evaluated with regard to its ability to solve the optimization problem. Depending on the representation a mapping of the chromosome, the genotype, to the actual solution, which is denoted as phenotype, is necessary. This genotype-phenotype mapping should avoid introducing a bias. For example, a biased mapping could map the majority of the genotype space to a small set of phenotypes. The genotype-phenotype mapping is not always required. For example, in continuous optimization, the genotype is the solution itself. But many other evolutionary modeling processes require this mapping.

An example is the evolution of machine learning pipelines. Each step in a machine learning pipeline can be coded as binary part in a genome, see Fig. 2.3 for a mapping from a bit string to `Python` commands from `sklearn`. For example, 10 at the beginning of the bit string causes a normalization preprocessing step while 00 at the end calls ridge regression. Such a mapping from genotypes to phenotypes is an essential part of the GENETIC ALGORITHM modeling process.

2.6 Fitness

In the fitness computation step the phenotype of a solution is evaluated on a fitness function. The fitness function measures the quality of the solutions the GENETIC ALGORITHM has generated. The design of the fitness function is part of the modeling process of the whole optimization approach. The practitioner can have an influence on design choices of the fitness function and thus guide the search. For example, the fitness of infeasible solutions can be deteriorated like in the case of penalty functions, see Chap. 5. In case of multiple objectives that have to be optimized at the same time, the fitness function values of each single objective can be aggregated, for example by computing the weighted sum. This technique and further strategies to handle multiple objective functions at the same time are discussed in Chap. 6. An important aspect is a fair evaluation of the quality of a solution. It sounds simple to postulate that a worse solution should employ a worse fitness function value, but a closer look is often necessary. Should a solution that is very close to the global optimum, but constrained have a worse fitness value than a bad solution that is feasible? And should a solution that is close to the optimum of the first objective in multi-objective optimization, but far away from the optimum of a second objective, which is much less important, get a worse fitness function value than a solution that is less close to the first optimum but much closer to the second one? To summarize, the choice of the penalty for infeasible solutions and the choice of appropriate weights in multi-objective optimization are important design objectives.

Most approaches aim at minimizing the number of fitness function calls. The performance of a GENETIC ALGORITHM in solving a problem is usually measured in terms of the number of required fitness function evaluations until the optimum is found or approximated with a desired accuracy. Minimizing the number of fitness function calls is very important, if a call is expensive, for example, if a construction element has to be generated for each evaluation. Fitness function calls may also require a long time, for example, if a simulation model has to be run to evaluate the parameters generated with the GENETIC ALGORITHM. The machine learning pipeline that is evolved with a GENETIC ALGORITHM is a good example for a comparatively long fitness evaluation. For each evaluation the machine learning pipeline has to be trained on the data set. To avoid overfitting it is required to repeat the training multiple times with cross-validation, which additionally takes time. Finally, the accuracy of the prediction model has to be evaluated on a test set in order to get a precision score that can be used as fitness function value.

2.7 Selection

To allow convergence towards optimal solutions, the best offspring solutions have to be selected to be parents in the new parental population. A surplus of offspring solutions is generated and the best are selected to achieve a progress towards the optimum. This selection process is based on the fitness values in the population. In case of minimization problems low fitness values are preferred and vice versa in case of maximization problems. Minimization problems can easily be transformed into maximization problems with negation. Of course, this also works for transforming maximization problems into minimization problems.

Elitist selection operators select the best solutions of the offspring solutions as parents. Comma selection selects the μ best solutions from λ offspring solutions.

Plus selection selects the μ best solutions from λ offspring solutions and the μ old parents that led to their creation.

Many selection algorithms are based on randomness. Roulette wheel also known as fitness proportional selection selects parental solutions randomly with uniform distribution. The probability for being selected depends on the fitness of a solution. For this sake, the relative fitness of solutions normalized with the sum of all fitness values in a population, usually by division. This fraction of fitness can be understood as probability for a solution of being selected. The advantage of fitness-proportional selection operators is that each solution has a positive probability of being selected.

In case of comma selection good parents can be forgot. Also the randomness of fitness proportional selection allows forgetting of the best solutions. Although this might sound contra-productive for the optimization process at first, forgetting may be a reasonable strategy to overcome local optima. Another famous selection operator is tournament selection, where a set of solutions is selected randomly and within this competition subset, the best solutions are finally selected as new parents. The second step can be implemented with fitness proportional selection as typical example. Tournament selection offers a positive probability for each solution to survive, even if it has worse fitness values than other solutions.

When using selection as mechanism to choose the parents of the new generation, it is called survival selection. The selection operator determines, which solutions survive and which solutions die. This perspective directly implements Darwin's principle of survival of the fittest. But the introduced selection operators can also be employed for mating selection that is part of the crossover operators. Mating selection is a strategy to decide, which parents take part in the crossover process. It makes sense to consider other criteria for mating selection than for survival selection.

2.8 Termination

The termination condition defines, when the main evolutionary loop terminates. Often, the GENETIC ALGORITHM runs for a predefined number of generations. This can be reasonable in various experimental settings. Time and cost of fitness function evaluations may restrict the length of the optimization process. A further useful termination condition is convergence of the optimization process. When approximating the optimum, the progress of fitness function improvements may decrease significantly. If no significant process is observed, the evolutionary process stops. For example, when approximating the optima of continuous optimization problems, the definition of stagnation as repeated fitness difference lower than 10^{-8} in multiple successive generations is reasonable. Of course, stagnation can only mean that the search might have got stuck in local optima, hence missing the global one. Restart strategies, see Chap. 4, are approaches that avoid getting stuck in the same local optima. If the GENETIC ALGORITHM repeatedly approximates the same area in solution space although starting from different areas, the chance is high that the local optimum is a large attractor, and a better local optimum is unlikely to find. It can also be that this local optimum is the global one.

2.9 Experiments

The experiment has been the main analytic tool since the beginning of GENETIC ALGORITHM research. Hence, carefully conducted experiments have an important part to play. The first task before the experimental analysis is the formulation of a research question. As GENETIC ALGORITHM experiments have a stochastic outcome, a temptation of some researchers might be to bias the results by selecting only the best runs. However, a fair comparison shows all experiments, although one might feel that at least one run was bad luck. It may not have reached the optimum and thus may be disturbing the presentation of the average runs. To be statistically sound, at least 25 repetitions are usually required, 50 or 100 is also a recommendable choice. More runs are often not necessary, more than 1000 repetitions can already be bad as unlikely outliers might occur. In the extreme case of outstandingly expensive optimization runs 15, 10, or even 5 runs can be a necessary compromise.

Table 2.1 shows the experimental comparison between two GENETIC ALGO-RITHMS, a normal $(1+1)$-GENETIC ALGORITHM and a $(1+1)$-GENETIC ALGO-RITHM with nearest neighbor meta-model (MM-GA) on the two continuous benchmark functions Sphere and Rosenbrock [89]. The concept of a GENETIC ALGO-RITHM with meta-model, also called fitness function surrogate, will be introduced in Chap. 8. The experiments have been repeated 25 times. The results show the means and the standard deviations of all runs in terms of fitness function values after 5000 iterations. Iterations correspond to fitness function evaluations in case of a $(1+1)$-GENETIC ALGORITHM.

The results show that the GENETIC ALGORITHM with meta-model achieves lower fitness function values. The optimum of both benchmark functions lies at the origin with a fitness value of 0.0. The question comes up, if the algorithm that achieves a better fitness function value in average is really better in practice. Is the result resilient from a statistical perspective? An answer to this question is only possible,

Table 2.1 Experimental comparison of two GENETIC ALGORITHMS, one with, the other without fitness function meta-model on the Sphere function and on Rosenbrock, from [55]

Problem		$(1+1)$-GA		MM-GA		Wilcoxon
	d	Mean	Dev	Mean	Dev	p-value
Sphere	2	2.067e-173	0.0	2.003e-287	0.0	0.0076
	10	1.039e-53	1.800e-53	1.511e-62	2.618e-62	0.0076
Rosenbrock	2	0.260	0.447	8.091e-06	7.809e-06	0.0076
	10	0.519	0.301	2.143	2.783	0.313

if we perform a statistical test. It tells us, if the comparison between two algorithms is statistically valid.

The question for an appropriate statistical test is comparatively easy to answer. The famous student T-test is not applicable, since the outcome of GENETIC ALGORITHM experiments is not Gaussian distributed. But the Gaussian distribution is a necessary prerequisite of the T-test, which examines, if two sets of observations come from the same distribution. An appropriate test for GENETIC ALGORITHMS is the Wilcoxon rank sum test [104]. It does not make assumptions on the distributions of the data. Instead, the Wilcoxon test sorts the outcomes of both sets of experimental observations and performs an analysis solely based on the ranks of this sorting. A small Wilcoxon value of under 0.05 proofs statistical relevance. Coming back to Table 2.1, the results show that the GENETIC ALGORITHM with meta-model and a superior fitness is performing significantly better than its competitor, the simple GENETIC ALGORITHM.

2.10 Summary

GENETIC ALGORITHMS are successful optimization approaches that allow optimization in difficult solution spaces. In particular, if no derivatives are available and if the fitness landscape suffers from ill-conditioned parts, GENETIC ALGORITHMS are reasonable problem solvers. In this chapter we summarized the foundations of GENETIC ALGORITHMS. They are based on populations of solutions that approximate optima in the course of iterations. Genetic operators change the solutions. Crossover operators combine the genomes of two or more solutions. Mutation adds randomness to solutions and should be scalable, drift-less, and reach each location in solution space. The genotype or chromosome of a solution is mapped to a phenotype, the real solution, before it can be evaluated on the fitness function. The latter has to be carefully designed as it has a crucial impact on the search direction. Selection chooses the best solutions in a population for survival. These solutions are the parents of the following generation. With the introduced concepts at hand, we are already able to implement simple GENETIC ALGORITHMS. The next chapters will introduce useful mechanisms and extensions for GENETIC ALGORITHMS, which tweak their performance and make them applicable to a broad spectrum of problems.

Chapter 3
Parameters

3.1 Introduction

The success of GENETIC ALGORITHM optimization processes significantly depends on the choice of appropriate parameters. The question comes up how to find the optimal parameter choices. The problem is an optimization problem within the optimization challenge of the original problem. Moreover, some parameter setting and tuning tasks turn out to be dynamic optimization problems, as the optimal choice varies in the course of the optimization process. Some taxonomies differentiate between exogenous and endogenous parameters [22, 46, 51]. Exogenous parameters are global parameters of the GENETIC ALGORITHM defining global properties like population sizes and selection pressure. Endogenous parameters define properties on the level of chromosomes. The latter appear multiple times in a population and are usually excellent candidates for self-adaptive parameter control, which will be introduced in this chapter.

Parameter tuning and control techniques have been developed that allow tuning of GENETIC ALGORITHMS before the run. Settings can be tuned by systematically testing values, by employing latin hypercube designs, and by treating them as optimization challenges. Control strategies are designed for finding appropriate parameters during the run of an algorithm. Dynamic control strategies control the parameters depending on static schemes like the number of generations. Adaptive parameter control strategies use a feedback from the search like Rechenberg's mutation rate control. Self-adaptation is the automatic control of parameters based on a secondary genetic optimization process. Most parameter tuning and control strategies are broadly applicable. They can be implemented in most GENETIC ALGORITHM variants with only minor adaptations. With this chapter and the previous one, the depiction of the foundations of GENETIC ALGORITHMS is complete.

© Springer International Publishing AG 2017

O. Kramer, *Genetic Algorithm Essentials*, Studies in Computational Intelligence 679, DOI 10.1007/978-3-319-52156-5_3

3.2 Parameter Tuning

As first step in research on parameter tuning for GENETIC ALGORITHM search, many static settings were proposed. Particular parameter settings like the choice $\sigma = 0.1$ for mutation rates in bit flip mutation were usual. Later, it was discovered that no optimal settings exist that are feasible for all problems. This is part of the no-free-lunch problem that will be introduced in more detail in Chap. 7. There is no parameter choice that is optimal for all problems. Once we have found a good algorithm or parameter setting for one problem or a problem class, we can be sure that there are problems, on which the approach or particular parameter settings will fail.

Parameter tuning strategies treat the parameterization of GENETIC ALGORITHMS as optimization problem. There are various kinds of ways to tune parameters. Some are supported by statistics like latin hypercube sampling. The latter is a method that promises to cover the data space while sampling from a multidimensional distribution with as few samples as possible. It generates only one sample in each dimension.

Others are based on simple grid search. Unlike many machine learning approaches like support vector machines, the parameter tuning problem in GENETIC ALGO-RITHMS is more crucial to proper settings. For parameters like the regularization parameter and the width of the radial basis function kernel of support vector machines, a coarse grid search in the exponents of ten are often sufficient like the interval $[10^{-20}, 10^{-19}, \ldots, 10^{20}]$. Such a coarse tuning is not sufficient for GENETIC ALGO-RITHM settings, for example for mutation rates.

Expert knowledge is a good source for proper parameter settings. The practitioner has a good feeling for the behavior of the GENETIC ALGORITHM. If domain knowledge is available, guesses and estimations by the expert will be close to the true optimal parameter settings. Approaches exist that support the parameter calibration process with statistical methods [15].

3.3 Meta-Genetic Algorithm

The meta-GENETIC ALGORITHM uses a GENETIC ALGORITHM to tune the parameters of a GENETIC ALGORITHM that solves the actual optimization problem [16]. Algorithm 2 shows the pseudocode of the meta-GENETIC ALGORITHM. In the main generational loop a population of parameter candidates is produced with crossover and mutation. It is very similar to the basic GENETIC ALGORITHM introduced in Chap. 2 with the exception that the fitness computation is not conducted on a fitness function, but on the outcome of the repeated runs of an inner GENETIC ALGORITHM that solves the actual optimization problem. The inner GENETIC ALGORITHM has to be repeated multiple times as its outcome is stochastic. For the inner GENETIC ALGORITHM, a reasonable measure has to be computed that is basis of the outer GENETIC ALGORITHM's fitness function. Such a measure can be the median, mean, or the best run. It is also reasonable to take into account the standard deviation to be sure that an outlier is not responsible for the evaluation of the current parameter set.

Algorithm 2 Meta-GENETIC ALGORITHM

1: initialize population
2: **repeat**
3:   **repeat**
4:     crossover of parameters
5:     mutation of parameters
6:     run GENETIC ALGORITHM multiple times
       with parameters
7:   **until** population complete
8:   selection of parental parameters
9: **until** termination condition

The outer GENETIC ALGORITHM often achieves better parameter settings than the ones a practitioner could have chosen, but this performance has its price. The major disadvantage of the meta-GENETIC ALGORITHM is its inefficiency. The computational effort to achieve this objective is enormous. The inefficiency results from the fact that each fitness function evaluation of the outer GENETIC ALGORITHM causes multiple runs of the inner GENETIC ALGORITHM. Finally, the question comes up, if the investment of the complete load of fitness function evaluations into the parameter tuning process would better have been invested into the search in the original solution space.

3.4 Deterministic Control

In the remainder of this chapter we concentrate on parameter control strategies, which are designed for finding the best parameters during the genetic optimization run. In deterministic control an external scheme is used to control the parameters during the run [1, 26]. Coupling the mutation rate to the generation counter is a common strategy. It allows the adaptation of the mutation rates during the run. A multiplicative decrease like $\sigma' = \sigma \cdot 0.9$ can be applied. In most cases a reduction of the mutation rate is necessary to allow convergence towards the optimum. But the rigid behavior can be disadvantageous in many situations. The external scheme does usually not exactly match the optimal mutation rate decrease. Furthermore, an increase is not possible without a feedback from the search process. A fast optimization is only possible, if the mutation rate adaptation scheme exactly matches the requirements, which is also again a parameter tuning problem. Figure 3.1 illustrates a typical mutation rate decrease performing a linear development with an additive scheme like $\sigma' = \sigma - 0.01$. Two exemplary optimal mutation rates leading to the highest progress rates could be achieved with different control strategies. The dynamic control strategy does not match the optimal ones, but develops similarly. In situations that require a high flexibility in particular in situations, where an increase of mutation rates is advantageous, dynamic mutation rate control fails. The strategy cannot know how to react without feedback from the search.

Fig. 3.1 Illustration of
actual parameter control
(*solid line*) representing a
typical mutation rate
decrease, and two exemplary
optimal developments that
do not match, but that are
close to the control strategy

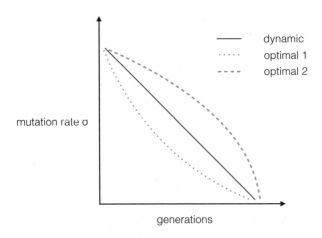

3.5 Rechenberg

Adaptive control strategies are designed for considering feedback from the search.
A famous mechanism for adaptive parameter control is Rechenberg's mutation rate
control [86]. Its idea is to increase the mutation rate whenever possible to accelerate
the search. Further, a decrease of the mutation rate is reasonable, if the search gets
stuck. Rechenberg's idea is to measure the success probability of the mutation operators. Algorithm 3 shows the pseudocode of the Rechenberg rule. If it is successful
with a probability higher than 1/5, the rule increases the mutation rate, usually in an
exponential way. For a (1 + 1)-GENETIC ALGORITHM success can easily be measured
via the ratio of successful generations. If we choose a reference number of generations, for example five oriented to the illustration of Fig. 3.2, and we observe that
the GENETIC ALGORITHM generates a better offspring solution than its parent in 3 of
the 5 generations, we assume that larger steps are possible in solution space to move
faster towards the optimum. An increase of the mutation rate can simply be achieved
by multiplying it with a factor larger than 1. The mutation rate is decreased, if the
success rate is lower than 1/5. It is not changed, if the success rate is equal to 1/5.
For example, when one solution is successful in five generations, no change of the
mutation rate is recommended.

　　The mechanism of Algorithm 3 has to be embedded within the generational loop
of a GENETIC ALGORITHM. The choices of the number of generations we observe
the search before we apply the Rechenberg rule and the factor we use to modify for
the multiplicative increase and decrease of the mutation rate define the speed of the
adaptation behavior.

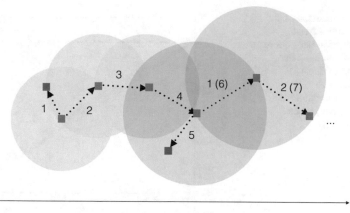

better fitness

Fig. 3.2 Illustration of Rechenberg's mutation rate control strategy

---

Algorithm 3 Rechenberg rule

1: measure success
2: **if** success rate $> 1/5$ **then**
3:    increase mutation rate
4: **else if** success rate $< 1/5$ **then**
5:    decrease mutation rate
6: **end if**

The idea of the Rechenberg rule is to stay in the so-called evolution window, which guarantees optimal progress rates. To illustrate the advantages of adaptive parameter control in continuous solution spaces, Fig. 3.3 shows a comparison between a simple $(1+1)$-Genetic Algorithm without mutation rate control and the same Genetic Algorithm with Rechenberg's mutation rate control on the Sphere function with 10 dimensions. Both Genetic Algorithms were run for 10,000 generations corresponding to the same number of fitness function evaluations. The fitness is plotted on a logarithmic scale. The plot shows the average, best, and worst of 30 runs. All other runs lie in the shadowed area. The results show that the adaptive mutation rate control is necessary to let the Genetic Algorithm approximate the optimum. The fitness development is linear on the logarithmic scale. Such approximation behavior is desirable in continuous solution spaces.

For Genetic Algorithms that use a population of candidate solutions the Rechenberg rule can also be applied. A reference like the best fitness of each generation can be used to evaluate the generation-wise success. If the offspring population is large enough, the rate of successful solutions can be measured and an adaptation of the mutation rate is possible in each generation.

Fig. 3.3 Comparison of a
GENETIC ALGORITHM
without mutation rate control
but constant mutation rate
(const.) and a GENETIC
ALGORITHM with
Rechenberg's mutation rate
control (Rechen.) on the
Sphere function

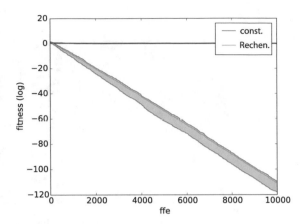

3.6 Self-adaptation

Self-adaptation in GENETIC ALGORITHMS is the automatic evolutionary control of
mutation rates [94]. In self-adaptive parameter control each solution gets an own
mutation rate that is subject to crossover and mutation. The mutation rate becomes
an endogenous parameter in contrast to the dynamically controlled mutation rate and
Rechenberg's strategy, where the mutation rate is a global exogenous one. As the
mutation rate is bound to each individual, it can be inherited with the selected solutions. Algorithm 4 shows the pseudocode of the self-adaptive GENETIC ALGORITHM.
It is closely related to the basic GENETIC ALGORITHM of Chap. 2, but employs strategy parameters that are subject to crossover and mutation, and are inherited over
generations bound to the candidate solutions.

For crossover, standard continuous crossover operators are usually applied. Mutation rates for continuous representations are usually positive real numbers, for example specifying the Gaussian distribution. For mutation of the mutation rates special
operators are available. Very appropriate for the self-adaptive operator of continuous mutation rates is the log-normal mutation operator. It allows an adaptation in the
exponents of the exponential function exp and a logarithmically linear approximation
of the optimum with

$$\sigma' = \sigma \cdot \exp(\tau \cdot \mathcal{N}(0, 1)). \tag{3.1}$$

The mutation rate τ serves as mutation rate of the mutation rates. There are recom-
mendations for the choice of τ that depend on the problem dimensionality [5]. The
optimal choice has been theoretically determined for simple functions, for example
for the unimodal and symmetric Sphere function.

Algorithm 4 Self-adaptive GENETIC ALGORITHM

1: initialize population of **x** and σ
2: **repeat**
3:   **repeat**
4:     crossover of σ
5:     crossover of **x**
6:     mutation of σ
7:     mutation of **x** with σ
8:     fitness computation
9:   **until** population complete
10:   selection of parental population
11: **until** termination condition

Self-adaptation can be extended in various kinds of ways. First, for multivariate Gaussian mutation, it can be extended for adapting multiple mutation rates at once. For this sake, an individual is equipped with a vector of mutation rates, one for each dimension. This is advantageous for difficult solution space conditions. To handle a mutation rate vector, the log-normal mutation operator is adapted. A global part applies Eq. 3.1 to all components of the mutation rate vector at once while an individual part applies the equation to each single mutation rate component. This can be described in one equation and implemented in few lines of source code. Two mutation parameters τ are introduced, one for the global and one for the component level.

As highly developed field, further extensions for parameter control in continuous solution spaces have been developed. One is correlated mutation that can be implemented with a diagonal matrix rotating the mutations or with a set of angles that describe the rotation for each axis. To allow self-adaptation, each individual must be equipped with such a rotation matrix or set of angles. Another alternative for parameter control in continuous solution spaces is self-adaptive biased mutation. Unlike the principle of drift-less mutations, self-adaptive biased mutation allows to bias the Gaussian distribution by shifting its center. Direction and magnitude control the bias.

Self-adaptation can also be applied to exogenous parameters like population sizes and selection pressures. This is achieved by equipping each individual with a parameter and aggregating these to a global one. The aggregation of endogenous parameters is a mapping of local individual-based parameters to one global parameter. It is similar to averaging all parameters by taking into account all parents for arithmetic crossover, which would result in one global aggregated parameter in the end.

Self-adaptation is the search in a space of parameters that is performed simultaneously during the main optimization process. The search in parameter space is guided by the success in the primary solution space. This is hardly possible, if the parameter search space is high-dimensional with only few candidate solutions. Moreover, self-adaptation may suffer from premature stagnation of the search process. A frequent phenomenon is that mutation rates decrease before reaching the optimum due to drastically decreasing success rates, for example in the vicinity of constraint boundaries. This will be discussed in Chap. 5.

3.7   Summary

The choice of appropriate parameters is essential for the success of GENETIC ALGO-
RITHMS. Static parameters that are kept constant during the GENETIC ALGORITHM
run can be tuned beforehand. Sampling strategies like the latin hypercube sampling
and grid search are often applied to tune parameters. Statistical methods can support
the tuning process. Some parameters must be controlled during the run for a significant improvement of the search. An easy parameter control technique is the dynamic
control with the typical linear or exponential decrease of parameters like mutation
rates during approximation of the optimum. A parameter control technique is much
more flexible when considering feedback from the search. This allows to flexibly
adapt to the search process. The Rechenberg rule is an example for mutation rate
control based on the success rate during the optimization process. It is an excellent
method for continuous mutation rate control and only requires the specification of
parameters that define the magnitude of the mutation rate change.

Self-adaptation of mutation rates lets the evolutionary process take control of the
parameter adaptation problem. Successful parameters are inherited with the individual and spread over the population. They are subject to crossover and mutation before
they are applied to the chromosome of the candidate solution. Self-adaptation can
also be used for discrete parameters.

Part II
Solution Spaces

Chapter 4
Multimodality

4.1 Introduction

Many optimization problems are difficult to solve. The existence of numerous local and global optima can significantly complicate the search. Local optima have a better fitness than their environment. Such a local optimum may be the global one, but it is often not. Further, the GENETIC ALGORITHM does not know, if it has already found the global optimum. The fitness landscape can be very hilly with many neighboring local optima that attract optimization algorithms like GENETIC ALGORITHMS. Algorithms that employ a population of candidate solutions may get trapped in areas of local optima, which we also call niches or basins in the following. Contrariwise, populations of candidate solutions allow a GENETIC ALGORITHM to get rid of local optima, as the chance increases with a larger number of candidate solutions to jump outside into another niche of the solution space.

Optimization problems with a fitness landscape employing many local optima are called multimodal. To illustrate the problem of numerous local optima, Fig. 4.1 shows the fitness landscape of a solution space with four local optima that are at the same time global optima. For many practical problems it can be desirable to detect as many local and potentially global optima as possible. For other problems it may be desirable to get rid of local optima and only concentrate on hunting the global one. For the situation illustrated in the figure it may be desirable to find all local optima. This offers the practitioner the possibility to choose among different solutions with different characteristics. For example, in engineering, such an alternative solution could require different materials. If there is a shortage of any kind of material type, an alternative solution might allow the flexibility to change the production process.

There are many standard mechanisms in GENETIC ALGORITHMS that aim at coping with local optima. A population targets at increasing the chance of exploring parts of the solution space apart from the majority of the existing solutions. Mutation operators provide the possibility for jumps outside of a niche by allowing large steps in solution space with a small but positive probability. Most selection operators

© Springer International Publishing AG 2017
O. Kramer, *Genetic Algorithm Essentials*, Studies in Computational
Intelligence 679, DOI 10.1007/978-3-319-52156-5_4

Fig. 4.1 Multimodal fitness
landscape with four optima

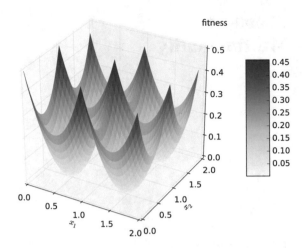

have a positive probability for worse solutions of being selected, although they are
outperformed by solutions in attracting niches.

Apart from the standard GENETIC ALGORITHM mechanisms like populations,
mutation, and selection operators, strategies have been proposed that explicitly aim
at taking care of multimodal solution spaces. Restart strategies repeat the search
multiple times, each time from different parts of the solution space. Fitness sharing
assigns the same fitness to numerous solutions in a niche to maintain diversity in
the population. Novelty strategies aims at exploring new parts of the solution space.
Niching strategies detect potential locations of optima and search in each niche until
the corresponding optima are approximated.

4.2 Restarts

A simple strategy against getting stuck in local optima and waste time in approximating a basin of the solution space that might not be the global one is to perform random
restarts. The GENETIC ALGORITHM is started multiple times, see the illustration in
Fig. 4.2. Starting from different parts in solution space, each optimization process
approximates the closest local optimum. It may be reasonable to change parameters
in each run. This change can include the starting position, but also population sizes,
mutation rates, and selection pressure. A famous strategy is to double the population
size for each new restart. This is applied in a restart variant of the CMA-ES for example [70]. Doubling the population size increases the chance of exploring different
parts of solution space. When reaching the same local optima in successive GENETIC
ALGORITHM runs, this is a good indicator that the current parameterization should
be adapted to a more explorative strategy with larger population sizes and mutation
rates.

Fig. 4.2 Illustration of
restart strategy. Starting from
different parts in solution
space each optimization
process approximates the
closest local optima

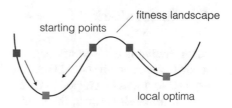

A perspective on restart strategies is that the GENETIC ALGORITHM performs
some kind of random walk in the space of local optima. Each run reaches one of the
closer local optima. But the GENETIC ALGORITHM does not exploit information of
neighboring local optima and instead simply restarts the evolutionary process from
an arbitrary point in solution space.

More focused on controlling the global search in the space of local optima is the
idea to globally control the mutation rate. For example, in [52] we combine local
search that is usually associated with detecting a local optimum in a basin of the
solution space with a mutation rate mechanism that acts on a global level. Algorithm 5
illustrates the local mutation rate control idea. It increases the global mutation rate,
if the same local optimum is repeatedly reached to achieve exploration. The search
is not restarted at a random position, but uses the last local optimum as basis for the
mutation with the global mutation rate. In turn, the mechanism decreases the mutation
rate on the global level, if new solutions are found after each local approximation.
The mechanism is similar to Rechenberg's success, but *reverses* its direction. The
decision, how often a local optimum must be reached until the mutation rate is
increased, is a performance tradeoff problem. The probability of being sure that
the same local optimum is reached, increases with the number of repetitions, which
might be expensive. The proper choice of the magnitude of the change of the mutation
rate depends on the solution space characteristics. The same holds for the threshold
that defines the similarity of solutions. For example, if we are not interested in local
optima that differ more than 10^{-2}, we should define this threshold accordingly and
we may be able to terminate the search in each local optimum after stagnation in the
range of 10^{-3}.

Algorithm 5 Global mutation rate control

1: **repeat**
2:     mutate local optimum
3:     local search
4:     **if** same local optimum **then**
5:         increase mutation rate
6:     **else**
7:         decrease mutation rate
8:     **end if**
9: **until** termination condition

4.3   Fitness Sharing

Fitness sharing is the idea of assigning the same fitness to numerous solutions in a niche to maintain diversity in the population. A simple but effective variant is to assign all solutions in a region to their average fitness value. The selection operator cannot distinguish between the solutions anymore and is forced to treat them in the same kind of way. This allows numerous worse solutions to survive and thus hinders the evolutionary process to converge too fast to optima in niches. Promising solutions can continue their walk in solution space. However, such a mechanism hinders convergence against attractive areas, which is the typical tradeoff between exploration and exploitation.

If solutions spread over a hilly fitness landscape, see Fig. 4.3, fitness sharing makes this region look like a plateau (dashed blue line). Fitness sharing directly interacts with the selection operator. Most selection operators like elitist selection and fitness proportional selection will randomly choose any of the solutions with shared fitness instead of concentrating on the local optima of the hilly region. This will allow the search leaving this region and moving into a novel attractive area like the larger basin on the right hand side.

Fitness sharing requires the specification of the candidate solutions that share the same fitness. This can simply be defined by a randomly chosen solution within a neighborhood. For this sake, the size of the neighborhood in solution space must be specified. Clustering, which will be explained in more detail later in this chapter in the context of niching strategies, is another possibility to identify groups of chromosomes that should share the same fitness function values. Clustering aims at identifying groups of solutions with similar characteristics.

It is important that the fitness sharing mechanism can be switched off in the course of the optimization process to allow convergence to potential optima. More advanced are mechanisms that adapt the fitness sharing mechanism. The combination with novelty search operators, see next section, will allow biasing the search faster away from the fitness plateaus. Within a region of the solution space that is protected against selection with fitness sharing, novelty search allows a walk towards unexplored parts.

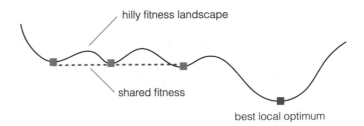

Fig. 4.3 Illustration of fitness sharing for avoiding getting stuck in hilly parts of the solution space

4.4 Novelty Search

Novelty search is an interesting technique that aims at exploring unknown areas of the solution space. Unknown areas may accommodate local or even global optima. If local optima attract the search, the exploration of unknown areas may allow the detection of potential basins useful to search in. Novelty search requires the evaluation of the novelty and uniqueness of solutions. Various measures are available for this evaluation. A reasonable measure is the distance of a solution to the solutions in a population or to the solutions in an archive that is managed during the search process. Various distance measures are reasonable in this context. Most often, the Euclidean distance is used.

An exemplary mechanism for novelty search for generating a candidate solution is presented in Algorithm 6. After initialization of the population new solutions are generated with crossover and mutation. The variant checks, if the new solution can be classified as outlier. If this is true, the novelty condition is fulfilled and the loop is exited. The rest of the GENETIC ALGORITHM works as usual. When the population is complete, it is evaluated on the fitness function for selection of the new parental population.

Algorithm 6 Novelty search operator

1: **repeat**
2:     crossover
3:     mutation
4:     outlier check
5: **until** novelty condition fulfilled

The novelty of a solution can also be evaluated on the prediction error of a meta-model of the fitness function. This idea is based on the assumption that a novel solution from an unknown part of the solution space results in a bad meta-model accuracy as the solution space is sparsely covered with solutions in that area. The meta-model or surrogate concept will be introduced in Chap. 8. It is a machine learning method that is trained on an archive of past solutions. The solutions play the role of patterns while the fitness function values are the corresponding labels. This novelty measure is not unproblematic. The prediction error can also be large in case of noise and highly structured data spaces. But also in such solution space areas, further search is reasonable.

4.5 Niching

Niching is a technique that separates GENETIC ALGORITHM optimization processes to focus on specific parts of the solution space. The idea of niching is that attractive basins in solution space exist that must first be identified. Then, independent evolutionary processes concentrate on optimization within these niches. Biological niches in nature allow species to share an environment without competing for the same resources. Avoiding competition results in more genetic diversity at one place. This also holds for the search with GENETIC ALGORITHMS. If there is no global selection pressure on the whole population, but only in niches, the search can maintain enough diversity to approximate local optima in these basins (Fig. 4.4).

Often, clustering methods are employed to detect niches in multimodal fitness landscapes. Clustering groups patterns according to intrinsic properties of the data. It belongs to the class of unsupervised learning methods in machine learning. Unsupervised methods do not employ label information for learning, but learn models solely based on the intrinsic structure of patterns. Densities and variances in data sets allow the identification of groups that belong together and consequently define a cluster. Numerous clustering methods have been introduced in literature. One of the most popular ones is k-means that iteratively repeats assigning solutions to the closest cluster centers and re-computing the cluster centers of all assigned solutions.

Within a niche, the search can be focused by adapting the mutation rate to a setting that makes it improbable to leave. From the perspective of the exploration and exploitation dilemma the operators of the GENETIC ALGORITHM care for exploring new promising regions in solution space while the clustering approach cares for the maintenance and therefore for the exploitation of explored regions, in which the convergence process can safely be conducted.

Figure 4.5 illustrates clustering with density-based spatial clustering of applications with noise (DBSCAN) and belongs to the density-based clustering methods [25]. With a user-defined radius, illustrated by the grey circles, and a minimum number of patterns, the density of points is estimated. If the density exceeds a certain threshold and more points lie within the radius, a point is classified as core point. All core points that lie within the radius of others belong to the same cluster. Points at the border that are no core points but lie within the radius of a core point also belong to that cluster. Patterns that are neither core points nor corner points are noise. Density clustering methods are attractive as they do not require an estimate of the number of clusters in the data sets and as they allow the detection of clusters of arbitrary shape. Instead, they depend on the specification of density properties. However, most density-based clustering methods suffer from the curse-of-dimensionality problem in high-dimensional data spaces. This is particularly disadvantageous for high-dimensional solution spaces. Further clustering methods that might be more appropriate to the particular solution spaces can be applied instead.

Fig. 4.4 Niching allows maintaining diversity in different basins of solution space, here for two niches

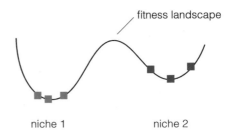

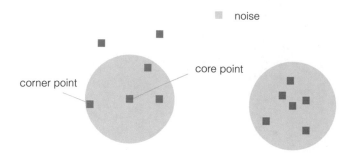

Fig. 4.5 Illustration of DBSCAN. Core points and corner points are defined via density parameters and result in clusters

4.6 Summary

Multimodal solution spaces are difficult challenges for optimization heuristics. The task is to approximate as many local optima as possible. This chapter gave an introduction to multimodal optimization methods. There are many strategies for detecting numerous local optima. Besides the standard mechanisms based on population sizes and parameter control, this chapter introduced numerous corresponding methods.

Restart strategies perform genetic search multiple times and can reach different local optima with the help of different starting points and conditions. With this strategy they try to approximate new local optima, which can potentially be the global ones. More focused on exploiting knowledge from the past search is the strategy to control a global mutation rate with the help of tracking successively reached local optima. Other techniques concentrate on maintaining diversity in a population. This can be achieved with fitness sharing that hinders selection from the reduction of diversity by assigning the same fitness to a set of solutions. Novelty search favors solutions that are different from the main population. Outlier detection methods can be used to test the novelty of a solution. Niching aims at explicitly detecting basins in the solution space and focusing the search in each niche. Niches are often detected with clustering algorithms. The techniques can also be combined to improve the success of finding as many local optima as possible. Niching strategies slow down the convergence process as they aim at maintaining diversity in the population. This tradeoff between exploration and exploitation is the usual price that has to be paid for dealing with multimodal optimization problems and algorithms. It is the task of the practitioner to carefully choose GENETIC ALGORITHM extensions like the ones presented in this chapter and to tune their relevant parameters in order to allow an effective and efficient optimization process. The development of methods that allow balancing the tradeoff between exploration and exploitation for certain classes of optimization problems is an active research field.

Chapter 5
Constraints

5.1 Introduction

Many practical optimization problems involve one or more constraints. Constraints reduce the solution space to a feasible subset. They can have numerous origins. Mathematical and logical restrictions, physical conditions like material constraints, and numerous further examples reduce a combinatorial or continuous solution space to a feasible subset. From the perspective of GENETIC ALGORITHMS a mechanism must be offered that allows treating the case that a solution is not feasible. This can be implemented with a constraint function. Similar to fitness functions the objective is to find the optimum with few constraint function calls. Different scenarios are possible. Constraint function calls might be cheap. In this case, a lot of sampling is possible to generate feasible solutions, which can be checked for their fitness in the second step. In case of expensive constraint function calls it is worth to spend effort on mechanisms that reduce their number significantly. Further, it is possible that we get different information about an infeasible solution. The solely information if a solution is feasible carries less information than detailed information about the magnitude of constraint violations of one or even more constraint functions.

For optimization with GENETIC ALGORITHMS the question comes up how constraints can be considered. One attempt is to adapt the genetic operators so that they can generate feasible solutions efficiently. In this chapter we introduce constraint handling techniques for GENETIC ALGORITHMS. The easiest one is death penalty that repeats generating solutions until a feasible one is available. Penalty functions deteriorate the fitness of constrained solutions to allow the search in infeasible parts of the solution space. Decoders map the constrained solution space to an alternative one that is not constrained or that employs easier constraints. There are numerous further constraint handing techniques, which are shortly overviewed in this chapter.

© Springer International Publishing AG 2017
O. Kramer, *Genetic Algorithm Essentials*, Studies in Computational
Intelligence 679, DOI 10.1007/978-3-319-52156-5_5

5.2 Constraints

Constraints reduce the feasible solution space to a smaller one (Fig. 5.1). In practice constraints can be logical restrictions, material characteristics, physical conditions, runtime bounds, and many more. The result can be that a solution is not applicable since it is not realizable, for example, a solution might crash a simulation model. For optimization purposes constraints are handled as functions that restrict the feasible solution space. For example, in continuous solution spaces, constraints can be formulated as equations and inequations. An equation can be expressed via two inequations with different signs. Without loss of generality, we can reverse the inequality by changing its sign. One can be transferred into the other by changing the sign. The result of a constraint function call can be a binary value indicating feasibility or infeasibility. But it can also be more detailed information giving insights into the magnitude of the constraint violation.

One way to handle constraints, which sounds simple at first, is to avoid them. The question is how to choose a representation and how do design genetic operators such that all restrictions are immediately fulfilled and all possible solutions are feasible. An example is the traveling salesman problem. A feasible solution starts at a city, visits all cities only once, and returns to the first at the end. If the solution to this problem is represented as list of cities, genetic operators might generate solutions that are not feasible. For example, the standard n-point crossover operator can generate solutions, which do not contain all cities or which visit the same city multiple times. The solution is the design of an operator that takes care of repeatedly visited and missing cities. Partially mapped crossover is a special recombination operator that has been designed to tackle this problem [32]. It changes a part of the genome of a combinatorial representation like the one of the traveling salesman problem and afterwards replaces all swaps that took place to avoid duplicates and restore deletions. The design of such feasibility preserving constraint handling methods can be a tedious task. A relatively simple variant for continuous solution spaces will be introduced in the following.

Fig. 5.1 Illustration of solution space with one linear constraint that divides the solution space into a feasible and an infeasible part

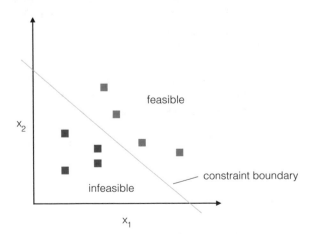

5.3 Death Penalty

One of the simplest methods to handle constraints is death penalty. Death penalty is sometimes referred to as penalty function, see the following section. But as it often significantly differs in its use and implementation, we devote it an own section. Death penalty belongs to the feasibility preserving constraint handling methods. Algorithm 7 shows the pseudocode of genetic operators that are forced by death penalty to generate feasible solutions. After crossover and mutation, the feasibility of the novel solution is checked. If the solution is not feasible, the process is repeated until its feasibility is fulfilled.

Algorithm 7 Death penalty

1: **repeat**
2:     crossover
3:     mutation
4:     check constraints
5: **until** solution is feasible

Death penalty is obviously an easy mechanism to handle constraints. However, there are good reasons not to use it. First, if the ratio of feasibility in the solution space is very low, death penalty is an inefficient method. Many tries may be required to generate one or even more feasible solutions. This tedious undertaking may be very inefficient in practice, in particular because the optimum often lies at the border or even in a corner of the feasible solution space part surrounded by constraints.

If the success rate is very low, meaning that it is very improbable to generate feasible solutions, a major problem is premature mutation rate reduction. With a mutation control mechanism like Rechenberg's success rule or self-adaptation, the mutation rates may drop due to decreasing success probabilities when employing death penalty. This phenomenon will be discussed later in this chapter.

5.4 Penalty Functions

A famous constraint handling approach is the reduction of the fitness of infeasible solutions with penalty functions. Penalty functions deteriorate the fitness of a constrained solution to make it less attractive for being selected for a new generation. In case of maximization problems the fitness is reduced. In turn, in minimization problems, the fitness is increased. The ratio of fitness changes depends on the amount of constraint violation, which may further be scaled by a penalty factor α. A constraint function $g(\mathbf{x})$ measures the constraint violation. The value is higher for larger constraint violations. Hence, in case of minimization problems, the fitness can be deteriorated but adding a positive value. A typical penalty function

$$f'(\mathbf{x}) = f(\mathbf{x}) + \alpha \cdot g(\mathbf{x}) \tag{5.1}$$

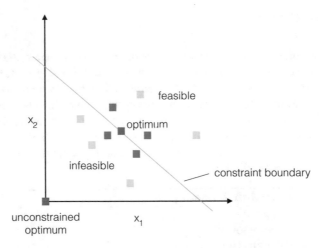

Fig. 5.2 Illustration of penalty function principle. It penalizes solutions in the infeasible part of the solution space and allows searching in the infeasible part of the solution space

implements this idea and scales the fitness deterioration with a continuous penalty factor α.

Penalty functions allow the search to take place in the infeasible part of the solution space. Figure 5.2 illustrates this situation. Darker solutions employ a higher fitness. The ones closer to the unconstrained optimum are penalized due to their constraint violation. Search near the border of feasibility is advantageous, as the optimum is often located at the border. Penalty functions can most efficiently be applied, if the magnitude of infeasibility can be measured with the constraint function yielding more than only a binary value. The penalty factor controls the balance between infeasible and feasible solutions by controlling the magnitude of the penalty. Penalty functions can only be applied, if the fitness of solutions can be measured or at least estimated although being infeasible.

As the choice of the penalty factor depends on the problem instance, there is no general recommendation. Instead, it may be reasonable to adapt the penalty factor in the course of the genetic optimization process [44]. Under the assumption that the optimum lies at the border of the feasible solution space, it is reasonable to balance the search. If the majority of the solutions is feasible, the penalty factor can be weakened to let the search take place in the infeasible part. In turn, the penalty factor can be increased, if too many solutions are infeasible. Such balancing approaches have been proposed recently in [61]. This approach implements the described mechanism by taking into account the feasibility of solutions during the last generations. It tries to keep the feasibility rate approximately around 1/5th, similar to the success rule of Rechenberg's mutation rate control, see Chap. 3.

Fig. 5.3 Illustration of repair approach. An infeasible solution is repaired by linear projecting with the help of the closest feasible solution or the fittest one

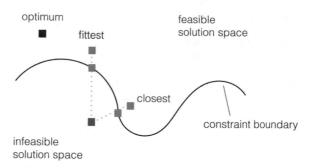

5.5 Repair

Repair approaches make infeasible solutions feasible. We mentioned the traveling salesman problem as example for a constrained problem at the beginning of the chapter. An infeasible solution would visit two or more cities multiple times or would not visit some cities at all. Such a solution could be repaired by removing duplicates or by adding missing ones. This can be done in a naive way by simple deletion and insertion processes. More sophisticated repair operators would consider the tour length when adding or removing cities.

For continuous solution spaces repair operators project an infeasible solution to a feasible one. Such a projection can be done by searching on a line between the infeasible solution and a feasible reference. Figure 5.3 illustrates the repair approach for a solution space with an infeasible solution and two solutions in the feasible region. The closest solution that is feasible is a good reference point for the projection process. The strategy to consider the tour length when repairing an infeasible traveling salesman tour is similar to considering the fittest solution in a population when repairing it. In continuous solution spaces a repaired solution can be based on the projection between the infeasible solution and the fittest solution, also see the illustration in Fig. 5.3. Repair approaches require knowledge about the solution space and have to be adapted to the employed representation. They might either be part of the genetic operators or of the generational loop.

5.6 Decoders

Decoders map the constrained solution space to an unconstrained one or at least to a solution space with less difficult conditions. The genetic optimization process takes place in the unconstrained space, which can make the optimization process much easier. The trick is that a mapping must be available that covers as much solution space as possible with similar characteristics. A mapping back from the decoder space is also necessary for getting the original chromosome encoding a solution to the problem.

Fig. 5.4 Decoder functions map a complicated constrained solution space to an easier one, where the search is less difficult to perform

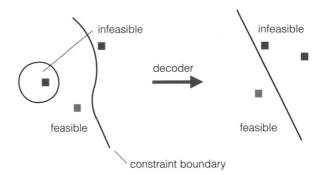

Figure 5.4 illustrates the concept of a decoder function. The original constrained solution space has a curved constraint boundary and holes. After the decoder function mapping the solution space is smoother without holes but with a linear constraint boundary. For the latter a linear constraint boundary meta-model can be learned, see Chap. 8.

The key question is how to design such a decoder function. This is surely a difficult challenge and significantly depends on the solution space. Kernel functions known from support vector machines [103] can be employed for this mapping. Kernel functions are mainly known in machine learning for the purpose of handling nonlinear data spaces. A mapping back from the kernel space to the original space can be a tedious task.

5.7   Premature Stagnation

A problem that often occurs in case of solution space conditions with low success probabilities is premature stagnation of the fitness function approximation process. Constrained problems may suffer from premature stagnation at the boundary of the feasible solution space. The reason is mainly a major decrease of mutation rates caused by low success probabilities.

Figure 5.5 illustrates the continuous situation for the Sphere function with one linear constraint. Here, premature stagnation can occur because the area of success, which is the area with better fitness, has disadvantageous properties. The closer the search comes to the optimum of the constrained problem, the less the direction of the optimum attracts the search. Contour lines of the fitness function, two are exemplarily shown in Fig. 5.5, lie almost parallel to the linear constraint when being close to the optimum. Hence, the constraint boundary attracts solutions that are no directed to the optimum. There are even solutions (dark blue) that are further away from the optimum but employ a better fitness than solutions that are closer with worse fitness (light blue). Further, the probability of generating feasible solutions when using death penalty as constraint handling method and a spherical mutation

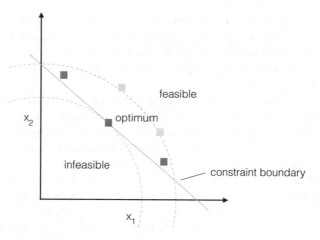

Fig. 5.5 Premature stagnation can occur at the constraint boundary in continuous solution spaces [50]

operator like Gaussian mutation, significantly decreases in the neighborhood of the constraint boundary.

To prevent premature stagnation in this specific case, numerous strategies can be employed. A mutation operator that is able to adapt to the shape of the constraints is advantageous, for example correlated mutation that can rotate an elliptical Gaussian shape to the angle of the constraint line [50, 94]. Further, a minimum mutation rate can prevent premature stagnation [50, 62]. To allow convergence to the optimum, a mechanism is required that reduces the minimum mutation rate in the course of the optimization process. Such a reduction mechanism can be based on the rate of infeasible solutions, which increases when coming closer to the constraint boundary and decreases, when the mutation rate is reduced.

5.8   Summary

Constraints can make a difficult optimization problem even more difficult to solve. In this chapter we introduced the constrained optimization problem for GENETIC ALGORITHMS. We discussed various definitions of constraints. The emphasis of the chapter was on introducing constraint handling techniques for GENETIC ALGO- RITHMS. One of the simplest and most successful mechanism is death penalty that discards infeasible solutions until enough feasible are available. Penalty functions decrease the fitness of solutions that are infeasible. They are easy to implement, easy to control and allow the search taking place in the infeasible region of solution space. With a controllable penalty factor, penalty functions can be adapted to put an emphasis on the infeasible part of the solution space.

Repair approaches repair infeasible solutions based on mechanisms that are adapted to the solution space characteristics, for example linear projections in case of continuous solution spaces. Decoder functions map the constrained solution space to a decoder space with simpler properties, where the search with GENETIC ALGO-RITHMS is easier to perform.

There are numerous further ways to handle constraints in GENETIC ALGORITHMS. An interesting one is the treatment of constraints as separate objectives that have to be taken into account. For this sake, evolutionary multi-objective optimization techniques can be used [68] that are introduced in the next chapter. They balance constraint violation and fitness function optimization. The result is a Pareto-front of solutions with different degrees of constraint violation and fitness. At the end of the optimization process the search concentrates on feasibility of the solutions.

A further interesting aspect in constraint handling is the reduction of the number of constraint function calls. This can be achieved with a constraint meta-model. Chapter 8 will introduce such a meta-model that can be used instead of the real constraint function to reduce the number of constraint function calls.

Chapter 6
Multiple Objectives

6.1 Introduction

Up to now we only considered single-objective problems. In practice many optimization problems involve two or more conflictive objectives. Conflictive means that when getting better in one objective, at least one other objective deteriorates. In such cases we call an optimization problem multi-objective. There are numerous examples for conflictive objectives in practice. A typical conflict occurs between cost efficiency and performance. The better the performance of a system is, the larger its costs usually become. A further typical conflict occurs between weight and stability. The lighter a system is the worse is its stability. If the practitioner is aware of the weights between objectives, this information can be used to transfer a problem with multiple objectives into a single-objective problem. In such case the resulting objective function is the sum of all weighted objective functions, for example $f = w \cdot f_1 + (1 - w) \cdot f_2$ in case of objective functions f_1 and f_2 and weight $w \in [0, 1]$ that balances between both objectives.

Without a decision for a weight of the objectives, it is difficult to solve the optimization problem. As no unambiguous comparison between solutions that are optimal concerning at least one objective is possible, solutions of that kind are incomparable. However, solutions that are worse in all objectives are outperformed and from this perspective useless. Once a weight has been chosen, the optimization problem can be treated as single objective problem. However, the challenge in multi-objective optimization is to approximate a set of solutions that constitute a compromise between all objectives. The goal becomes to evolve a set of solutions that are not dominated by other solutions meaning that they are not worse in all objectives. This set is also known as Pareto-set. The fitness values of the solutions in a Pareto-set build the Pareto-front. As GENETIC ALGORITHMS are based on populations of solutions, the evolution of Pareto-sets is naturally possible.

The key in evolving a Pareto-front with multi-objective GENETIC ALGORITHMS is the selection operator. Most selection operators are based on two steps. The first step is usually non-dominated sorting that sorts the solutions according to the level of

© Springer International Publishing AG 2017
O. Kramer, *Genetic Algorithm Essentials*, Studies in Computational
Intelligence 679, DOI 10.1007/978-3-319-52156-5_6

Fig. 6.1 With two
objectives a solution divides
the objective space into four
quadrants, here for a
minimization problem

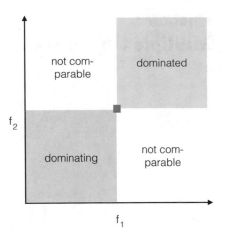

domination. The second step builds upon the non-dominated solutions and optimizes
a secondary criterion, mostly targeting at spreading solutions across the Pareto-front.
This chapter will introduce three examples for such secondary selection criteria. Non-
dominated sorting maximizes the Manhattan distance of the neighboring solutions.
Rake selection spans parallel lines in objective space and selects the closest solutions to each rake. Hypervolume-based selection maximizes the dominated objective
space. All three secondary mechanisms aim at achieving a good coverage of solutions
on the Pareto-front (Fig. 6.1).

6.2   Multi-objective Optimization

Multi-objective optimization is the problem of optimizing two or more conflictive
objectives at a time. Conflictive means that getting better in one objective usually
results in getting worse in another one. This leads to the definition of optimality
in multi-objective optimization. A solution is Pareto-optimal, if it is not dominated
by any other solution in solution space. Figure 6.2 shows how one solution divides
the objective space into four quadrants. The lower left quadrant contains solutions
that dominate the solution we consider. All solutions in this quadrant are better with
regard to both objectives. Solutions in the upper right quadrant are dominated by the
considered solution. The upper left and the lower right quadrant contain solutions
are not comparable. The set of non-dominated solutions consists of solutions that lie
in such quadrants with regard to all other solutions.

The set of non-dominated solutions corresponds to a subset of solutions in decision
space. The Pareto-set is the set of non-dominated solutions with regard to the whole
solution space and corresponds to the optimum in single objective optimization. The
fitness values of the solutions in the Pareto-set build a Pareto-front in objective space.
Most figures in this chapter illustrate a typical Pareto-front. Neither the Pareto-set

Fig. 6.2 Non-dominated
sorting sorts all solutions
with regard to their
non-domination rank. The
non-dominated solutions are
closest to the Pareto-front in
comparison to solutions of
higher rank

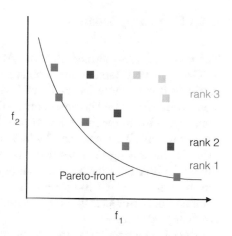

nor the Pareto-front have to be connected. In solution space different areas can exist
that result in neighboring positions on the Pareto-front. Further, for positions on the
Pareto-front numerous alternative solutions may exist in objective space. Niching
techniques can be applied for identification of such equivalent Pareto-subsets [57].
Like in single-objective multimodal optimization, the advantage of approximating
many optimal, near-optimal, or Pareto-optimal solutions is that the practitioner can
choose among alternatives, if a favored solution becomes infeasible.

6.3 Non-dominated Sorting

Non-dominated sorting [17] is the first step of most multi-objective optimization
approaches like NSGA-II, rake selection, and selection based on the hypervolume
indicator. It sorts the GENETIC ALGORITHM's population with regard to the non-
domination rank. Each solution that is not dominated by another solution belongs to
the first rank. If the solutions of the first rank are removed, the second rank consists of
solutions that are not dominated anymore. This process is continued until the set of
solutions is empty. Figure 6.2 shows an illustration of solutions belonging to different
ranks. The blue squares represent solutions that are not dominated. The dark grey
squares belong to the second rank of non-dominated solutions, the light grey to the
third rank. The non-dominated solutions of the first ranks are basis of a secondary
selection process.

Secondary criteria that push the solutions towards the Pareto-front and that aim
at achieving a broad coverage of solutions on the Pareto-front are introduced in the
remainder of this chapter. Often, more solutions are required for the secondary step
than the first subset of non-dominated solutions employs. In such cases it is a common
practice to take solutions from the second, third, or even higher ranks.

6.4 Crowding Distance

The non-dominated sorting GENETIC ALGORITHM, which is commonly known as NSGA, is a very famous multi-objective GENETIC ALGORITHM. The successor NSGA-II is even more successful and will be introduced in the following. As first step, NSGA-II is based on non-dominated sorting. Among the non-dominated solutions or a union of the first ranks of non-dominated solutions, NSGA-II seeks for a broad coverage. This is achieved with the crowding distance, see Fig. 6.3. It selects the solutions among the ones that have been selected via non-dominated sorting. Figure 6.3 illustrates the optimization objective of NSGA-II. For each solution the crowding distance, which corresponds to the Manhattan distance of two neighboring solutions for two objectives, is computed. The ones with the largest crowing distance are finally selected.

Instead of computing the two closest solutions on both sides of each solution and then computing their Manhattan distance, an alternative approach is presented in the original paper introduced by Deb et al. [17]. Algorithm 8 illustrates the approach. The idea is to sort the population with regard to each objective. For each solution it sums up the fitness difference of the two neighbors in this sorting. The first and the last solution in this sorting are assigned to an infinite crowding distance and are thus immediately selected.

Algorithm 8 NSGA-II

1: **for** each objective **do**
2:     sort population with regard to objective
3:     take first and last solution
4:     **for** each solution **do**
5:         add distance between left and right neighbors
6:     **end for**
7: **end for**

Fig. 6.3 Illustration of crowding distance, which is the Manhattan distance between left and right neighboring solution for two objectives

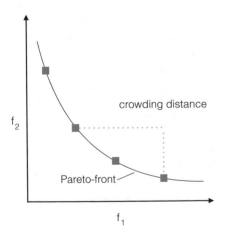

In case the practitioner has weights in mind for each objective the multi-objective problem can be translated into a single-objective one as pointed out in the introduction. This shrinks the area of interest to a line in objective space defined by the weights. But it still might be desirable to approximate the Pareto-front while having the weights in mind. Friedrich et al. [29] introduced an approach that allows the combination of weights with the crowding distances. They propose to multiply the crowding distance with the weights in the NSGA-II crowding distance computation. In Algorithm 8 this comes into play in the last step. For each objective and for each solution the distance between the left and the right neighbors is added weighted with the corresponding objective weight. The result of this procedure is a Pareto-front with a bias towards the objectives with larger weights.

6.5 Rakes

A straightforward approach of maintaining a uniform spread of solutions in objective space is rake selection. Similar to NSGA-II it is first based on non-dominated sorting. Among the solutions of the first non-domination rank, the ones are selected that are closest to parallel lines in objective space. These lines are placed equidistantly employing a rake base as connection between the extreme solutions with the best fitness for each objective.

Figure 6.4 shows an illustration of rake selection. Algorithm 9 presents the corresponding pseudocode. The rakes are placed orthogonally on the rake base and thus reach into the area of the objective space, where non-dominated solutions lie. For each rake the closest point among the solutions with rank one is selected. If the number of non-dominated solutions is too small, solutions from the next rank can take part in this selection process. The distances between points and lines can efficiently be computed regardless of the objective space dimensions.

The optimal points that are basis of the rake base can be computed first by optimizing each single objective and second by computing the further coordinate points

Fig. 6.4 Illustration of rake selection. The rake base connects the optimal solutions with regard to the single objectives. Rakes are placed orthogonally to the rake base, often equidistantly. For each rake the closest solution is selected

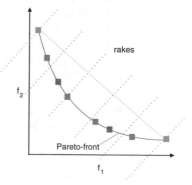

in objective space with the fitness for the remaining objectives. Alternatively, the rake base can be computed with the best solutions with regard to each objective of the current population. This on-the-fly computation of the rake base will result in varying positioning of rakes, but allows their flexible adaptation during the optimization process.

Algorithm 9 Rake selection

1: generation of population
2: non-dominated sorting
3: computation of rake base
4: placement of rakes
5: selection with regard to each rake

The rake approach works best for Pareto-fronts that have a linear shape. The more curved the front is, the less parallel it is cut by the rakes. An adaptation of the rakes according to the shape of the current set of non-dominated solutions is possible. A similar mechanism has been introduced in [58], where a ratio is computed to adapt the rake distances in each step based on the connection between the rakes and selected solutions.

6.6   Hypervolume Indicator

A prominent approach to approximate the Pareto-front is maximization of the hypervolume indicator that is also known as S-metric [4]. The hypervolume indicator measures the area in solution space that is dominated by a population. Considering a reference point that is dominated by all solutions, the dominated hypervolume can be computed.

Figure 6.5 shows the dominated objective space of a whole population. The grey reference solution in the upper right part of the objective space serves as opposite corner of the rectangles and therefore as reference for the computation of the dominated area. The overlap of rectangles has to be considered for a correct area computation. For more than two objectives the resulting space is a hypervolume. Maximizing the hypervolume has two effects. First, it leads to a good coverage of the non-dominated solutions for a broad spread among the approximated Pareto-front. Second, it pushes the solutions towards the real Pareto-front as maximization of the hypervolume leads to a movement away from the dominated reference point.

The computation of the hypervolume can become a complicated undertaking for more than two objectives. In particular in case of many objectives, the volumes often overlap and are not easy to compute. Moreover, the selection of the best solutions that maximize the hypervolume is a combinatorial optimization problem. The task is to find the subset of solutions, for which the metric is maximized. To avoid this, GENETIC ALGORITHMS with hypervolume indicator employ a $(\mu + 1)$ population scheme. In each generation only one solution is generated, the contribution of each solution to the overall hypervolume is computed and the one is discarded with the

Fig. 6.5 Illustration of a
GENETIC ALGORITHM that
maximizes the dominated
hypervolume in objective
space. For the hypervolume
computation a dominated
reference solution is required

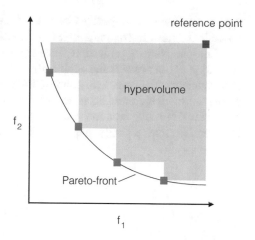

least volume contribution. Figure 6.5 highlights the contribution of one solution to
the hypervolume with the light blue area as part of the overall hypervolume.

6.7 Summary

Many practical optimization problems involve the optimization of more than one
objective. If the objectives are conflictive, not all solutions are comparable. If a solution is worse in all objectives, it is dominated. Solutions that are non-dominated are
interesting for the practitioner, as they approximate the Pareto-set. The counterpart
of the Pareto-set in objective space is the Pareto-front. The set of non-dominated
solutions is interesting when searching for the Pareto-set. GENETIC ALGORITHMS
are excellent methods for approximating multi-objective problems as they are based
on solution sets.

Non-dominated sorting is part of most multi-objective GENETIC ALGORITHMS.
It assigns each solution to a rank of domination and pushes the search towards the
Pareto-front. A secondary selection criterion is applied afterwards to improve the
spread among the Pareto-front. NSGA-II maximizes the space between solutions
on the Pareto-front to avoid agglomerations. Rake selection minimizes the distance
between solutions and lines that are equidistantly distributed in objective space.
The GENETIC ALGORITHM with hypervolume indicator maximizes the dominated
objective space and pushes the solutions towards the Pareto-front, which is only
indirectly the case for NSGA-II and rake selection. For the latter mechanisms can
also be implemented that maximize the distance to a dominated reference point.

Further, a hybridization between the different mechanisms is possible. For example, half of the solutions could be selected according to the rakes in solution space
while the other half can be based on maximizing the hypervolume. A hybrid may
profit from the different characteristics of the employed methods. Many further

approaches for multi-objective GENETIC ALGORITHMS and variants have been proposed in the past. Examples are NSGA-3 [107] and SPEA [96].

Multi-objective optimization finds many applications, for example in robotics [18]. A further interesting application is balancing of machine learning methods, in particular model complexity and runtime versus accuracy. On the one hand side, prediction models should be as accurate as possible, on the other hand, they should not suffer from long runtime and should not be too complex to avoid overfitting. Chapter 9 presents a similar approach for balancing machine learning ensembles.

Part III
Advanced Concepts

Chapter 7
Theory

7.1 Introduction

In the early decades of GENETIC ALGORITHMS theoretical investigations were seen as less important or even questionable. But for a consistent expertise in GENETIC ALGORITHMS, expert knowledge from practical investigations should be complemented by theoretical results. The theoretical analysis of GENETIC ALGORITHMS can significantly contribute to the understanding of their practical behavior. Today, a collection of theoretical results and tools has been introduced to support and extend the important knowledge of the practitioner.

The theoretical analysis of GENETIC ALGORITHMS mainly concentrates on the two properties runtime and convergence. Runtime analysis makes statements about the time algorithms need for solving a particular problem class. It neglects constants and thus focuses on complexity classes. Convergence analysis puts a focus on the analysis, how good an algorithm is able to approximate the optimal solution. The peculiarities of GENETIC ALGORITHMS are their iterative improvement of solutions and their randomness, which complicate the theoretical analysis.

This chapter gives an introduction to theoretical concepts with examples. It starts with an introduction of the proof technique of fitness-based partitions for runtime analysis. Then it shows how Markov chains can be used to analyze the convergence properties of populations. The behavior of GENETIC ALGORITHMS in continuous solution spaces can be characterized with progress rate analysis. Progress rates describe the movement towards the optimum with regard to the actual generation. The no free lunch theorem states that there is no superior algorithm or parameterization for every problem, but there is always a problem, for which a particular GENETIC ALGORITHM will fail. The schema theorem analyzes the behavior of short, low-order chromosome parts and their behavior within evolutionary processes. The chapter closes with a short introduction to the building block hypothesis, which aims at explaining the behavior of crossover operators when dealing with such schemata.

© Springer International Publishing AG 2017
O. Kramer, *Genetic Algorithm Essentials*, Studies in Computational
Intelligence 679, DOI 10.1007/978-3-319-52156-5_7

7.2   Runtime Analysis

Runtime analysis focuses on the runtime of algorithms on specific problems depending on their size. The latter is usually measured via the length of the input fed into the solving algorithm. For example, in case of the traveling salesman problem, the problem size is the number of cities the salesman has to visit. In case of bit string problems the problem size corresponds to the bit string length. Runtime analysis is a powerful tool in theoretical computer science. For deterministic algorithms runtimes can be guaranteed, in which the optimal solutions are found. In case of the employment of randomness this is not directly possible. But theory can still prove expected average and worst case behaviors.

The fitness-based partitions proof gives an upper bound on the average runtime. An example for its successful application is the runtime analysis of a $(1+1)$-GENETIC ALGORITHM on the problem OneMax. The optimization problem is to maximize the number of ones in a bit string, which is obviously a quite simple optimization problem, but interesting from a theoretical perspective. The $(1+1)$-GENETIC ALGORITHM is only employing bit flip mutation and accepts the better of two solutions in each generation. The runtime analysis aims at proving an upper bound of $O(N \log N)$ on the average runtime. This is better than quadratic or exponential runtime of many optimization approaches. The fitness-based partitions proof can be applied, if the solution space can be divided into sets of solutions with same or similar fitness. These sets are called partitions.

Figure 7.1 illustrates the concept of a partitioned solution space. All solutions with the same fitness belong to a partition that is represented as blue or grey rectangle. The partitions are sorted from low fitness at the bottom to high fitness at the top. In case of the OneMax problem each partition contains the bit strings with the same number of ones. Hence, $N+1$ partitions exist, if the length of the bit string is N. The probability to leave for a partition with higher fitness is the probability for flipping one of the zeros to a one while not flipping a one back to zero. The reciprocal of this probability is the expected number of generations, as it corresponds to a Bernoulli experiment. With the harmonic series we get the runtime $O(N \log N)$.

The key principle of partition-based proofs is to find a partitioning of the solution space with as few partitions as possible and a high probability to leave the current

Fig. 7.1 Fitness-based partitions divide the solution space into disjoint sets of equal fitness. For the proof the probability for mutating into any partition with higher fitness has to be determined

partition for any partition with better fitness. The larger the probability to jump into higher partitions the lower is the corresponding expected number of steps for success. Numerous further tools have been introduced for evolutionary runtime analysis, of which many have been introduced in [80], recently applied, for example in [19] for a runtime analysis of a $(1+\lambda)$-GENETIC ALGORITHM on OneMax. For the latter Gießen and Witt [30] analyze the interactions of population sizes and mutation rates.

7.3 Markov Chains

Many theoretical analyses for GENETIC ALGORITHMS focus on the behavior of populations with Markov chains, which are tools that treat the population at one generation as state. Markov chains are a general tool for analyzing stochastic processes. To model the stochastic influences of genetic operators, each state is described by probabilities to transfer into a successive state called transition probability. A matrix is used to describe the probabilities for transitions to any other state given the current state. In a canonical GENETIC ALGORITHM each state only depends on the individuals in binary representation [91]. Random changes of the genes in a population that are caused by the genetic operators are modeled in a transition matrix. Each state transition corresponds to a matrix multiplication. The transition matrix is the product of stochastic matrices each modeling crossover, mutation, and selection, respectively. Hence, it is possible to model the convergence of the whole population. An important property for Markov chain analysis is the independence of predecessor states and the assumption of a finite number of states. Such a Markov chain is called homogeneous.

Eiben et al. [21] studied the convergence properties of GENETIC ALGORITHMS as one of the first early works in this field. They showed that under certain conditions, a GENETIC ALGORITHM optimizing a function over an arbitrary finite space converges to an optimum with probability one. Rudolph [91] comes to the conclusion that convergence to the global optimum is no inherent property of a canonical GENETIC ALGORITHM without elitism. Instead, with elitist selection the convergence is guaranteed. Although Markov chain analysis leads to a deeper understanding of GENETIC ALGORITHMS, its applicability is restricted to simple cases as the resulting transition matrices are large growing with the dimensionality of the problem.

7.4 Progress Rates

The progress rate analysis has been developed for GENETIC ALGORITHMS in continuous solution spaces like evolution strategies [5, 6]. Progress rates are local performance measures evaluating the amelioration power of GENETIC ALGORITHMS. If the progress is measured in terms of fitness values, it is called progress gain [5]. We consider the progress rate in solution space. It is defined as the expected distance of the population center from the optimum. Figure 7.2 illustrates the movement towards

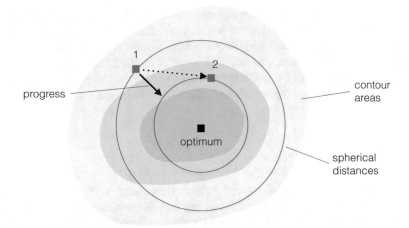

Fig. 7.2 The progress rate is defined as the expected movement towards the optimum in solution space, here exemplarily shown for one optimization step

the optimum in solution space for one step of a GENETIC ALGORITHM. The progress rate depends on numerous factors, for example on the problem with its fitness function and on the GENETIC ALGORITHM type. In many theoretical investigations the progress rate can only be determined for comparatively simple fitness functions.

Most progress rate analyses focus on the Sphere function, which is symmetric and its fitness values only depend on the distance to the optimum. Beyer and Schwefel [5] summarize some results based on progress rate analysis. For example, GENETIC ALGORITHMS with plus selection always have positive progress, which is independent of the chosen mutation rates. They continuously converge to the optimum. Further, GENETIC ALGORITHMS with comma selection can converge, if the mutation rate is chosen appropriately. This is a result that we assume in practice when facing parameter adaptation problems. If the mutation rate is chosen too large, the GENETIC ALGORITHM shows divergent behavior moving away from the optimum. Another result is that the progress rate increases with an offspring population larger than one. This is a good argument for employing populations instead of simple GENETIC ALGORITHMS that are only based on one parent and one child. Further, a population-based GENETIC ALGORITHM for continuous solution spaces with plus selection always performs better than the corresponding GENETIC ALGORITHM with comma selection. The best performance is achieved with arithmetic crossover employing the whole parental population as parents. Interestingly, GENETIC ALGORITHMS with crossover improve their performance using higher mutation rates. Further results are available for noisy problems. For the Sphere function with noise the $(1+1)$-GENETIC ALGORITHM degrades increasingly with the noise magnitude. In contrast, multi-recombination strategies perform well.

7.5 No Free Lunch

There ain't no such thing as a free lunch. This statement was adapted into a theory on optimization strategies by Wolpert and Macready [105]. Generally speaking, the no free lunch theorem states that there is no overall superior optimization algorithm that is able to solve all kinds of optimization problems. Nothing is free means that an algorithm that is adapted to a certain problem class and specific problem instances, where it performs considerably well, will be outperformed by other algorithms on other problems. The no free lunch theorem is an impossibility theorem. Originally, it assumes that all problems are equally likely inputs to the GENETIC ALGORITHMS. The consequence is that all GENETIC ALGORITHMS approximately show the same performance over all objective functions. However, as the theoretically possible number of strategies is limited in practice, it is actually possible to develop algorithms that perform best, at least for practical problem classes. A further argument that is frequently discussed in the context of no free lunch is the reevaluation of solutions. The better performance of a GENETIC ALGORITHM, which reevaluates a solution in comparison to a GENETIC ALGORITHM that does not, can be independent of the specialization to the particular optimization problem.

The no free lunch theorem can hardly be adapted to natural evolution. This is because fitness spaces in nature are structured by laws of nature while the theoretically possible number of problems and fitness functions is not restricted. But this independence of structure is a necessity in the reasoning of the no free lunch theorem.

7.6 Schema Theorem

John Holland introduced the famous schema theorem for the analysis of GENETIC ALGORITHMS [39, 40]. The schema theorem analyzes the proportion of schemata, which are candidate solutions in bit string representation with wildcards. Hence, a schema represents a set of solutions, which coincide at the bit positions, where no wildcards are. Schemata also known as building blocks define hyperplanes in solution space. The schema theorem states that short above-average fitness schemata in solutions spread in the population with higher probability. This is illustrated in Fig. 7.3, where the blue schema spreads in the population over the course of generations. The increase of such schemata during the genetic optimization process is also known as genetic drift.

The proportion of individuals representing a schema at subsequent time steps is given by the product of its probability of being selected and the counter probability of being disrupted. The probability of the disruption of a schema can be computed taking into account the probability for crossover multiplied with the probability that one-point crossover chooses a location within the end points of the schema. With such a definition solutions with high fitness get more copies in the course of the evolutionary process while below-average strings get few copies. Interestingly, the

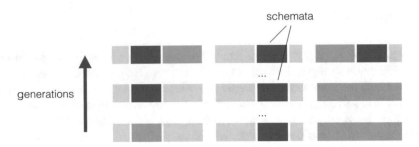

Fig. 7.3 According to the schema theorem short above-average fitness schemata spread in the course of the genetic optimization process

schema theorem only considers the disruptive effects of mutation and crossover. Although diversity is achieved with mutation, the mutation probability must be low to weaken its disruptive effect.

Some criticism about the schema theorem has been introduced by Holland himself. An analysis on the so-called royal road functions has shown that bad solution fragments may be inherited in early phases of the optimization process while being bound to good schemata and the consequently high fitness of the overall solution. This observation is also called hitchhiking, as the bad solution parts participate in the spread while not being responsible for the success.

7.7   Building Block Hypothesis

The building block hypothesis tries to explain the meaning of crossover. While the schema theorem only considers the disruptive effects of crossover, the building block hypothesis focuses on its constructive effects. Numerous experiments have shown that crossover and mutation work well in practice. Goldberg [31] and Holland [40] developed the building block hypothesis, which claims that crossover combines short, low-order schemata with high fitness to increasingly fit offspring solutions. These schemata, now called building blocks, are combined with crossover in the course of the genetic optimization process.

In the nineties Forrest and Mitchell [28] introduced an experimental setting for the building block hypothesis and came to the conclusion that a simple hill-climbing algorithm is faster than a GENETIC ALGORITHM with crossover on problems that fit the conditions of the building block hypothesis. Later, Jansen and Wegener [43] introduced a function, for which it can be shown that crossover is advantageous. Figure 7.4 illustrates the function based on bit strings with an optimum at $x^* = (1, \ldots, 1)$. A GENETIC ALGORITHM finds many bit strings with $N - k$ and also with k ones. For crossover probabilities lower than $1/N$ the runtime is $O(N^2 \log N)$ while mutation alone requires $O(N^k)$ generations. The interesting question comes up, if practical problems will be identified in the future, for which a theoretical result for the benefit of crossover can be found.

Fig. 7.4 Illustration of a fitness function, for which it has been proven that crossover can improve the runtime

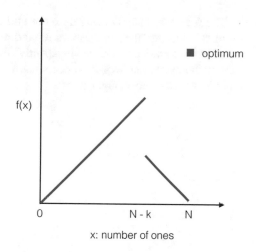

7.8 Summary

Theoretical investigations help the practitioner with solid knowledge that supports practical experiences. Various theoretical tools for GENETIC ALGORITHMS have been developed in the last decades. Theoretical results may be difficult to obtain despite their simplicity. A long line of research is sometimes required for proofs that look simple at the end. Often, theoretical results are only possible for simple algorithms and simple problems. Furthermore, many statements may be comparatively general and less precise than necessary for practical applications, for example runtime results with large constants. Nevertheless, they pave the way for a deeper understanding of GENETIC ALGORITHMS.

A theoretical analysis for many discrete and combinatorial problems is possible with a set of runtime tools. One of the simplest is the fitness-based partitions proof we introduced for the OneMax problem. Markov chains allow the analysis of convergence properties, if the solution space can conveniently be modeled with finite populations. To describe the convergence locally in the course of the evolutionary process, progress rate analysis is an adequate tool, which is in particular far developed for continuous GENETIC ALGORITHMS. Locality is a frequent assumption for many heuristics and proof principles.

Progress rate analysis has revealed many interesting results for the continuous Sphere function. The no free lunch theorem shows that no GENETIC ALGORITHM performs best on all problems, although no free lunch loses importance for the class of problems that can be computed in practice. The schema theorem analyzes the spread of low-order solutions during the optimization process. The constructive effect of crossover operators within the schema theorem is analyzed in the building block hypothesis. For crossover examples from runtime analysis demonstrate its success on artificial functions.

The future will surely bring up more and more theoretical results in the area of GENETIC ALGORITHMS and also for related randomized search heuristics. This may even lead to a unified theoretical perspective on most existing probabilistic optimization methods with an impact on decisions for the choice of methods, mechanisms, and the design of novel algorithms.

Chapter 8
Machine Learning

8.1 Introduction

Machine learning is the discipline of learning from data and observations. It combines statistics and learning paradigms from artificial intelligence. This chapter introduces concepts to support GENETIC ALGORITHMS with machine learning. For a detailed introduction to this field see [56]. Machine learning evolved to a very successful area of research in the last decades. It can mainly be divided into the two parts supervised and unsupervised learning. Supervised learning means learning from data with labels. Labels are additional information available for some training data. The task is usually to predict them for unknown data. If the labels are binary or discrete, the learning task is a classification problem. If labels are continuous, the task is called regression problem.

Unsupervised learning means learning without labels, but exclusively from the structure of the data itself. Clustering and dimensionality reduction are two variants of unsupervised learning. In the past numerous examples have shown that machine learning provides excellent tools to support GENETIC ALGORITHMS reaching from covariance matrix estimation to visualization of optimization runs with dimensionality reduction.

First, this chapter will introduce the concept of covariance matrix estimation for adapting the Gaussian distribution in continuous solution spaces. Then, it will present supervised learning models for replacing the fitness function during the course of evolution and also the constraint function in constrained optimization. To visualize high-dimensional optimization processes, dimensionality reduction can be employed. Chapter 9 will give an example for the reverse direction, which is the application of GENETIC ALGORITHMS to machine learning problems.

© Springer International Publishing AG 2017
O. Kramer, *Genetic Algorithm Essentials*, Studies in Computational
Intelligence 679, DOI 10.1007/978-3-319-52156-5_8

8.2   Covariance Matrix Estimation

For continuous solution spaces we introduced the Gaussian mutation in Chap. 2. It makes the reasonable assumption that chromosomes in solution space are Gaussian distributed. But the distribution often differs from a symmetric isotropic Gaussian shape. Instead, it might be scaled in some direction corresponding to separate variances in each dimension. Moreover, such a scaled Gaussian distribution might even be rotated resulting in correlations between the dimensions. Mathematically, this can be described with a covariance matrix, which consists of entries that describe the correlation between any two variables with covariances. If such a covariance matrix, or more specifically its decomposition, is multiplied with isotropic Gaussian distributed numbers, the results are random numbers that are distributed according to a scaled and rotated Gaussian distribution. In statistics numerous covariance matrix estimation techniques have been proposed. Empirical covariance matrix estimation is a famous instance. We introduced a variant of GENETIC ALGORITHMS with Ledoit-Wolf covariance matrix estimation and analyzed its performance on a set of benchmark problems [54].

Figure 8.1 shows an illustration of the covariance matrix estimated during an optimization run of a continuous GENETIC ALGORITHM on the Sphere function. The plot shows the contour lines of equal fitness corresponding to contour lines of equal probability when sampling from the Gaussian distribution. The spherical conditions can clearly be observed. Such an estimate is helpful in continuous optimization processes.

GENETIC ALGORITHMS with covariance matrix estimation belong to the class of estimation of distribution algorithms [65], which are based on iteratively sampling from a distribution, selecting the fittest candidate solutions, and estimating the new distribution. The covariance matrix adaptation evolution strategy (CMA-ES) [36] is a popular variant. It does not estimate the covariance matrix directly, but approximates it during evolution. Besides the covariance matrix estimation, the CMA-ES also employs a derandomized mutation rate update rule, which is called cumulative step size adaptation [37, 82]. Numerous variants of the CMA-ES have been proposed in

Fig. 8.1 Illustration of a covariance matrix estimated during the optimization run of a continuous GENETIC ALGORITHM on the Sphere function, oriented to [54]

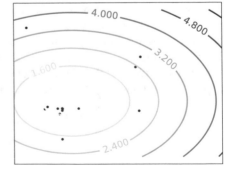

the past, for example a computationally efficient variant for limited memory [71] and a variant capable of handling noisy objective functions [63].

Algorithm 10 Covariance matrix estimation

1: initialize population
2: initialize **C**
3: **repeat**
4:   **repeat**
5:     crossover
6:     mutation with **C**
7:     fitness computation
8:   **until** population complete
9:   selection of parents
10:   estimation of **C** from parents
11: **until** termination condition

Algorithm 10 shows how covariance matrix estimation can be integrated into a GENETIC ALGORITHM. After initialization of the population and of the covariance matrix **C**, the main evolutionary loop generates new candidate solutions via crossover, mutation, and fitness computation. The mutation operator uses the covariance matrix that is estimated based on the selected parental population for the following generation. Covariance matrix estimation uses a set of the best recent solutions. This set can simply be the current parental solution. To save computational time, some approaches approximate the covariance matrix with an update rule performed every generation like the CMA-ES.

8.3 Fitness Surrogates

One of the main interests in optimization is to find the optimum with a minimum number of fitness function evaluations. Each evaluation might be expensive and a significant cost reduction can be achieved, if their number is minimized. This can even be more important for the practitioner than short runtimes. If we want to minimize the number of fitness function evaluations while time and computation are no limiting factors, we can learn a supervised machine learning model of the fitness function. This model can serve as surrogate that replaces the original fitness function. The model has the task to be a good estimator for fitness function values based on evaluations from the past. The tasks include the ability to interpolate the fitness function values in areas, where the solution is surrounded by other chromosomes, but also to extrapolate the fitness function values in areas, where no evaluation has been observed yet.

Figure 8.2 illustrates the fitness meta-model principle. The fitness function represented as solid line is evaluated for each solution represented as blue squares. The meta-model estimate is represented as grey dotted line. It almost fits to the actual

Fig. 8.2 Illustration of meta-model principle. The *horizontal axis* represents the solution space, the *vertical axis* illustrates the fitness function values

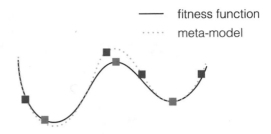

fitness function curvature with small deviations and is used to replace fitness function evaluations in case of the grey squares, where the predictions nearly match the real function values.

A machine learning model $\hat{f}$ is trained to predict the fitness function values f of novel solutions. Training is an optimization process for the model parameters. A candidate solution serves as pattern, its fitness as label. Training means that the parameters of the model are adapted such that they minimize the empirical risk, which is the deviation of the model predictions and the true labels. Such a minimization is only possible, if labeled data is available. The risk for overfitting the model to the training data is high, when the error is minimized on the training set. To avoid this, cross-validation divides the training set of labeled data into training and validation set. The training set is used to feed the model with patterns and labels, the validation set is used to examine the quality of the model on independent data. Based on the labels we know and the prediction of the training set, we can measure the model quality.

Algorithm 11 Meta-model operator

1: generate candidate solution
2: evaluate solution on $\hat{f}$
3: **if** solution fitness surrogate $\hat{f}$ fulfills quality criterion **then**
4:     evaluate solution fitness on f
5:     put evaluation into training set
6: **else**
7:     discard solution
8: **end if**

A meta-model management strategy is necessary for the decision, when to use it, to plan model updates, to tune parameters, and to plan exploration steps. In the course of the evolutionary process all solutions and their fitness evaluations are stored in an archive. This archive serves as training set. The learning model predicts the fitness of a new candidate solution. Algorithm 11 shows a genetic operator based on the meta-model. The solution is evaluated on the real fitness function, if the meta-model predicts a fitness that fulfills a certain quality, otherwise, it is discarded. In [55], we compare the fitness prediction to the fitness of the k-th best solution in the population.

If its fitness is predicted to be equal or better, it is worth to be tested on the real fitness function. Each evaluation on the real fitness function f will be put into the archive. Depending on the difficulty of the prediction problem and the employed machine learning algorithm, a new training and tuning of the model with grid-search and cross-validation might be necessary.

Numerous prediction methods have been proposed in the past. Famous ones are linear regression and nearest neighbor regression. Also support vector regression [92], the regression variant of support vector machines, have proven to be good surrogate models [72]. Support vector machines will be introduced in the next section.

8.4 Constraint Surrogates

In Chap. 5 constraints have been introduced. They reduce the allowed solution space to a feasible subset. A solution can be feasible or infeasible. From the perspective of meta-models, this can be treated as binary classification problem. Similar to the archive of fitness function calls introduced in the previous section, an archive of constraint function calls is managed that serves as training set. For the constraint surrogate similar mechanism have to be employed like for fitness function surrogates. Constraint surrogates can be used to pre-evaluate a solution.

Similar to regression methods, numerous classification methods are available in literature. Prominent examples are support vector machines [8, 92], which place a linear decision boundary in data space in a way that it correctly separates patterns of different classes and at the same time maximizes the distance of the patterns to this boundary. This distance is known as margin. Figure 8.3 illustrates support vector machine-based classification. The optimization problem to maximize the margin while patterns of different classes lie on different sides of the boundary is a constrained optimization problem. It can efficiently be solved with convex optimization

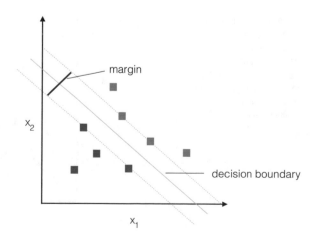

Fig. 8.3 Illustration of support vector machine-based classification. The support vector machine maximizes the margin while fulfilling the constraint that patterns of different classes lie on the correct side of the decision boundary

methods. Kernel functions and slack variables allow handling classification problems that are not linearly separable.

Our experiments have shown that the employment of constraint surrogates does not lead to a reduction of fitness function evaluations. Instead, it significantly reduces the number of constraint function calls while keeping the fitness function calls constant. When the constraints deliver continuous values as constraint violations, a regression model can be applied similar to the fitness surrogate case of the previous section.

8.5 Dimensionality Reduction for Visualization

GENETIC ALGORITHM optimization processes may take place in high-dimensional solution spaces, which cannot be visualized anymore. Besides the possibility to visualize only two or three dimensions at once, high-dimensional solution spaces can be mapped to two or three dimensions with dimensionality reduction approaches. Dimensionality reduction algorithms compute a set of low-dimensional counterparts of high-dimensional solutions without losing essential information. The maintenance of high-dimensional properties mainly concerns distances and neighborhoods. In other words, patterns that are neighboring in high-dimensional space should be neighboring in low-dimensional space as well. Moreover, patterns that are close to each other in high-dimensional space should be close to each other in low-dimensional space, and vice versa for patterns that are far away from each other.

Numerous dimensionality reduction methods exist. Principal component analysis (PCA) [45, 83] is a dimensionality reduction approach for linear data. It detects the axes in the data that employ the highest variances. Figure 8.4 illustrates the PCA concept. The projections of the patterns onto these axes, which are the principal components, are the novel low-dimensional representations. They can efficiently be

Fig. 8.4 Illustration of PCA-based dimensionality reduction. PCA identifies the axis in the data with the highest variances and computes the projections onto these principal components

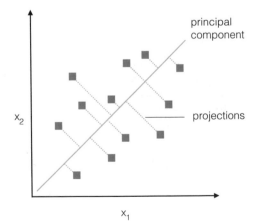

Fig. 8.5 Example for
dimensionality
reduction-based visualization
of a high-dimensional
GENETIC ALGORITHM run
with isometric mapping

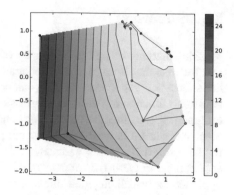

computed based on the covariance matrix of the data and the eigenvectors with the
largest eigenvalues. Moreover, nonlinear dimensionality reduction methods allow the
mapping of high-dimensional nonlinear data to low-dimensional spaces. Isometric
mapping [100] and locally linear embedding [90] are methods for nonlinear data.

Algorithm 8.5 shows the pseudocode of the dimensionality reduction-based visualization approach. First, the GENETIC ALGORITHM runs on the optimization problem as usual. An archive of candidate solutions that should be visualized is managed
during optimization. Afterwards, the dimensionality reduction approach maps the
population to a two-dimensional space. This process is also known as embedding.
To colorize the low-dimensional space, a mesh-grid of points within the convex hull
of embedded solutions is computed. A prediction model is trained with the embedded
points as patterns and the corresponding fitness values as labels to interpolate the
fitness of the points within this convex hull. If colors are employed, an interpolated
contour plot is the corresponding result. Last, the best solutions can be connected to
illustrate the movement of the evolutionary process. In [59], we employed isometric
mapping for the visualization process.

Algorithm 12 Visualization

1: run GENETIC ALGORITHM
2: embed population
3: compute convex hull
4: generate mesh-grid
5: interpolate contour plot
6: connect best solutions

Figure 8.5 shows an example for solution space visualization using the approach
of Algorithm 12 with isometric mapping. The figure visualizes the (1 + 1)-GENETIC
ALGORITHM run of Chap. 2. The plot shows the embeddings of the last 20 solutions
on the 10-dimensional Sphere function. The spherical contour lines of the Sphere
function are clearly visible. The lines follow the course of the optimization process.

8.6 Summary

Machine learning comprises a rich set of methods for learning prediction models and for mapping from high-dimensional to low-dimensional spaces. This chapter has shown how supervised learning can be used to reduce the number of fitness and constraint function evaluations. Regression models are applied for fitness function evaluations, classification models for constraint functions. A meta-model management mechanism is necessary for saving function evaluations and for organizing model updates and tuning parameters. Also unsupervised learning approaches find applications in GENETIC ALGORITHMS. We presented a dimensionality reduction approach for visualizing high-dimensional solution spaces with isometric mapping.

The niching approach introduced in Chap. 4 is a successful example for the application of clustering in GENETIC ALGORITHMS. After an initial sampling of the solution space and the selection of the best solutions, the clustering approach identifies potential basins that are locations of local optima. Many interesting hybridizations between GENETIC ALGORITHMS and machine learning will be subject to future research.

Chapter 9
Applications

9.1 Introduction

Numerous applications demonstrate the success of GENETIC ALGORITHMS. In this
chapter we show various examples from different domains. An overwhelming number of applications showed the strength of GENETIC ALGORITHMS in convenient
modeling, easy implementation, and efficient problem solving in the past. New and
improved applications are regularly presented at GENETIC ALGORITHM conferences
like the Genetic and Evolutionary Computation Conference (GECCO), the Congress
on Evolutionary Computation (CEC), and EvoStar, which have special devoted tracks
for this purpose. Recent applications of GENETIC ALGORITHMS include the biomedical domain [12, 78], arts [14], architecture [2], music [95], games [67, 69], the
energy domain [10, 87], engineering [99], and machine learning applications [79,
85].

This chapter presents various examples for the successful application of GENETIC
ALGORITHMS in different domains. It starts with the application of GENETIC ALGORITHMS to machine learning problems like unsupervised regression for dimensionality reduction. Evolutionary multi-objective optimization is used to balance ensembles
of classifiers. GENETIC ALGORITHMS are excellent methods for selection and tuning of features in prediction models. A corresponding example for feature tuning in
wind power prediction is presented. Wind turbines are more efficient, if wake effects
are avoided. We use GENETIC ALGORITHMS to find optimal turbine positions that
fulfill geographical constraints while maximizing the power output. Last, GENETIC
ALGORITHMS are used to optimize the control rules of virtual power plants.

© Springer International Publishing AG 2017
O. Kramer, *Genetic Algorithm Essentials*, Studies in Computational
Intelligence 679, DOI 10.1007/978-3-319-52156-5_9

9.2   Unsupervised Regression

An interesting line of research is the optimization of machine learning models with
GENETIC ALGORITHMS. They can be used for tuning the parameters of machine learning methods, but can also serve as main optimization methods for machine learning models. For example, GENETIC ALGORITHMS have successfully been applied
in balancing support vector machines [77] and for tuning dimensionality reduction
techniques [73]. Dimensionality reduction methods compute a mapping from high-
dimensional patterns to low-dimensional points. For a comprehensive introduction
to dimensionality reduction we refer to the depiction by Lee and Verleysen [66]. The
results can be used for visualizing high-dimensional data sets or as preprocessing
step for supervised learning methods like classification and regression.

Among the huge variety of dimensionality reduction methods unsupervised
regression is a very promising one [76]. The concept of unsupervised regression is
based on the idea of mapping low-dimensional points, which are the embeddings we
are seeking for, to the high-dimensional observed patterns. Many regression methods
can be used within this setting, but methods that allow mapping more than one label
at once are most appropriate. Nearest neighbor regression and kernel regression, also
known as Nadaraya-Watson estimation, allow the mapping to multiple labels. The
optimization problem is the minimization of the error between this mapping and the
original patterns. This is also referred to as data space reconstruction error, as the
patterns in data space have to be reconstructed by the mapping from low-dimensional
space to the high-dimensional one. This optimization problem is quite difficult to
solve. Figure 9.1 shows that the placement of only one point in low-dimensional space
is already a multimodal optimization problem. The figure visualizes the data space
reconstruction error when embedding a pattern into an existing solution on a sample
test data set [53] with nearest neighbor regression. For the optimization of numerous
patterns at once the fitness landscape becomes far more multimodal with many local
optima. As a complete solution consists of the coordinates of all low-dimensional
representations, the dimensionality of the optimization problem increases with the
number of patterns. When employing kernel regression the optimization problem can
be solved with gradient descent, as the derivative of the data space reconstruction

Fig. 9.1 Solution space of
unsupervised regression
when embedding one pattern
on a sample test data set

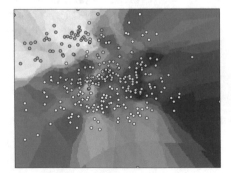

error with the Nadaraya-Watson estimator is analytically derivable. However, due to the numerous local optima the employment of GENETIC ALGORITHMS is an attractive approach. In [73], we analyzed the use of GENETIC ALGORITHMS with Gaussian mutation and mutation rate control to optimize unsupervised regression learning. It turns out that the combination of gradient descent with GENETIC ALGORITHMS, for example by alternating both optimization approaches, leads to the best results. GENETIC ALGORITHMS overcome local optima, gradient descent approximates the basins of attraction in solution space.

9.3 Balancing Ensembles

Ensembles of machine learning methods are famous for their ability to outperform their pure counterparts. The idea is also known as wisdom of the crowd. The predictions from numerous classifiers are combined to one single prediction. For classification and regression methods, but also for unsupervised algorithms and even optimization approaches, the hybridization turns out to be an effective way for improving accuracies. A key property of an ensemble is its diversity. The accuracy profits from varying training sets and parameterizations of the involved single classifiers. The drawback of numerous classifiers is the computational overhead that has to be managed. The most important objective in supervised learning is accuracy. Of course, this also holds for ensembles of classifiers. But another important objective is the computation time. Both objectives are obviously conflictive resulting in a multi-objective optimization problem. The number of ensemble members of each type and their parameters are the variables of the optimization problem. The parameters are bound constrained, as an upper bound on the number of classifiers is reasonable from a computational perspective and also the parameters space of machine learning methods is usually restricted. For discrete optimization variables NSGA-II with random resetting or Gaussian mutation with rounding can be applied.

Balancing ensembles of classifiers can be applied to a huge variety of ensemble variants. The experimental results in [81] show that multi-objective GENETIC ALGORITHMS are excellent approaches to evolve Pareto-fronts of ensemble classifiers. Table 9.1 shows an exemplary result of this work for an artificial benchmark classification data set generated with make_classification from the

Table 9.1 Error and corresponding runtime of nearest neighbor ensembles and random forests for minimizing error and for minimizing runtime, oriented to [81]

Ensemble	Minimum error		Minimum runtime	
	Error	Runtime	Error	Runtime
Nearest neighbors	0.240	24.4	0.419	0.006
Random forests	0.160	40.2	0.441	0.009

sklearn machine learning library [84] for nearest neighbors ensembles and for random forests [9]. The results show the classification error in terms of root mean square error and the ensemble runtime in seconds after multi-objective optimization with NSGA-II. The left part of the table shows the results for the ensembles that achieve a minimum error, the right part shows the corresponding results for the fastest ensembles. The figures show the price that has to be paid for the best accuracy in terms of runtime. In turn, the fastest ensembles only achieve a bad accuracy. A Pareto-set of ensembles is very useful in practical scenarios for the choice of an alternative a posteriori, i.e., after the optimization process.

9.4 Feature Tuning

For the integration of wind into the power grid the precise prediction of wind energy has an important part to play. Besides numerical models that simulate the atmosphere, models based on data have proven their success in the recent past. In [102] we could show that for the same location and timespan, the regression-based prediction outperforms the classic numerical models for short time horizons.

The idea of data-based models is to measure the wind power in the environment of a target turbine and to employ the regression techniques for mapping from the current wind power situation to the future. Figure 9.2 illustrates the situation. The possible time horizon depends on the temporal and spatial resolution of the wind power data. The wind speed must be related to the spatial dimensions of the surrounding turbines as well as to the temporal resolution of the time series data. The question comes up, which surrounding turbines are the best ones for the wind prediction problem, in particular with regard to the employed regression method. In [101] we use nearest neighbor regression as machine learning model.

To optimize the influence of the particular turbines in the environment, we employ GENETIC ALGORITHMS. Each solution is a vector of weights with the dimensionality of the number of turbines the wind prediction model makes use of. This weight vector scales the components of the wind time series patterns for the regression-

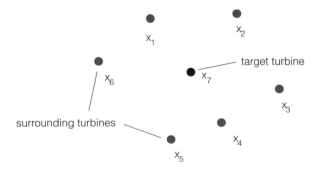

Fig. 9.2 The data-driven wind power prediction model is based on the time series data of the wind turbines that surround a target turbine, for which the prediction is computed

Table 9.2 Wind feature tuning of data-driven predictions with GENETIC ALGORITHMS, taken from [101]

Park	Persistence	GA start	GA final
Tehachapi	21.58	22.95 ± 0.67	16.45 ± 0.01
Reno	36.41	35.33 ± 1.15	27.24 ± 0.06

based prediction. Such a feature tuning is a common way in machine learning for helping regression and classification models focusing on specific dimensions. The GENETIC ALGORITHM that solves this continuous optimization problem is based on Gaussian mutation and Rechenberg's mutation rate control. It is a bound constraint optimization problem, as negative weights are not allowed. The bound restrictions are handled with death penalty.

In [101] we could show that the feature tuning process improves the prediction results significantly. In particular, for regression models that have no mechanism for tuning the feature importance like nearest neighbor regression, the GENETIC ALGORITHM optimization of weights yields large improvements in comparison to the untuned counterparts. Table 9.2 shows exemplary results for turbines near Tehachapi in California and Reno in Nevada for one-hour ahead predictions. The persistence model is a benchmark for comparison based on the assumption that the wind does not change within the prediction horizon. The prediction model is based on nearest neighbor regression employing 12 surrounding turbines with three past measurements. The table shows that the prediction model has a similar quality like persistence at the start of the optimization process. After 100 generations of the GENETIC ALGORITHM significantly better wind power prediction accuracies were achieved. The integration of numerous meteorological information can further improve the prediction accuracies.

9.5 Wind Turbine Placement

To increase the efficiency of wind turbines in wind farms, it is important to consider wake effects. Wake effects occur, when turbines stand in front of each other with regard to the wind direction. By reducing the kinetic energy of air molecules and by inducing turbulences, the wind power behind a turbine is decreased. This effect can be modeled in different kinds of ways in a simulation. We employ the model by Kusiak and Song [64] for modeling wake effects. It is based on the wind distribution that models frequencies and magnitudes of wind speeds from different directions. We use wind distribution data from the German Weather Service for locations in the northern part of Germany in our analysis, for example in [74]. The resulting energies achieved with the wind speeds that arrive at the turbines are computed with the power curves of real wind turbines. All these computations are required for the computation of the power output of a whole wind farm, which finally represents the fitness function for a GENETIC ALGORITHM in the wind turbine placement optimization scenario.

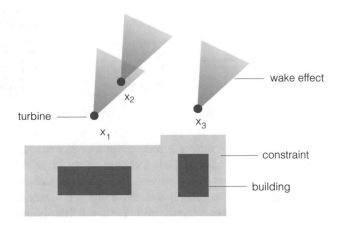

Fig. 9.3 Exemplary area of wind turbines with wake effects and building constraints. On the *left*, a turbine stands in the area of the wake effect of a neighboring one degrading its power output

Figure 9.3 shows an exemplary area of wind turbines with wake effects that reduce the power generation of neighboring turbines. A solution to the wind turbine placement problem is a vector of positions for all wind turbines. It turns out to be a constrained optimization problem as turbines have a least distance to other turbines and to buildings, streets, woods, and further geographical phenomenons. For a realistic scenario we use geo-information from OpenStreetMap [35], which contains coordinates and important properties for our wind park modeling. Figure 9.3 also illustrates the concepts of restricted areas induced by buildings.

The turbine placement problem can be treated as constrained continuous optimization problem. The application of a GENETIC ALGORITHM with Gaussian mutation is a good choice and turns out to deliver satisfying results in the end. We tested different operators and population sizes in [74] and later also experimented with various penalty functions to handle constraints [75]. The evolutionary tuning of the turbine positions significantly improves the power output of a wind farm. The GENETIC ALGORITHM places them in rows orthogonally to the main wind direction. The advantage in comparison to a manual placement is that some turbines are placed in configurations that would probably not be found without GENETIC ALGORITHMS. The turbine placement problem can be extended by taking into account arbitrary additional conditions. Administrative conditions or the preferences of customers can simply be considered within the constraint functions. The model also allows handling different kinds of turbines, in particular from different manufacturers and devices with different sizes. Related work on the use of GENETIC ALGORITHMS for turbine placement without constraints [33] and for large numbers of wind turbines [24] has also been published recently.

9.6 Virtual Power Plants

The integration of renewable energy resources into the power grid is a tedious task because of their fluctuations. For a stable grid the fluctuations of varying wind and solar power have to be compensated. The concept of virtual power plants is to bundle multiple different power resources to a single unit that fulfills certain properties. Virtual power plants often employ a fast conventional power plant that is able to compensate fluctuations and a power storage that is able to save over-capacities. In case of sudden power dropouts and undersupply the storage can inject saved energy while the control power plant has to increase its power generation in case of an empty storage. In case of over-capacities of power the storage can be charged. With one or multiple power consumers the overall consumption of power should be equal to the overall production.

The optimization problem is to minimize the absolute value of power in the system by controlling the actors, which are reserve power plants and storages, with a rule base. The rule base is optimized with GENETIC ALGORITHMS, an approach known as learning classifier systems. The rule base contains rules that control actors depending on the current system states. The strategy consists of rules with conditional parts that are selected according to the system states and action parts, see Fig. 9.4 that shows an exemplary rule base with a small set of rules. The system states depend on properties like wind speed, solar power, and storage load. The action part controls the actors of the system focusing on the battery and the reserve power plant. In [60] we model a solution as sequence of loading and injection steps for a storage facility and for powering up and down the reserve power plant.

The learning classifier system variables consist of discrete and continuous variables. The conditional part of each rule is randomly generated. With a minimum number of rules the space of system states will be sufficiently covered. The action part of the rule base is evolved with the GENETIC ALGORITHM. The evolved control strategies allow the virtual power plant to act flexibly and to balance power consumption and generation. Virtual power plants will play an increasing role in the future.

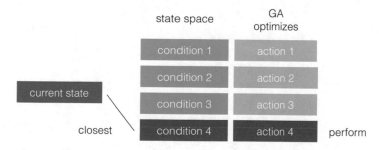

Fig. 9.4 Rule base of virtual power plant with four rules consisting of conditional and action parts. The conditional parts are randomly distributed in state space. The GENETIC ALGORITHM optimizes the action part. In each step the rule is performed that is closest to the current system state

Besides the effort of modeling learning classifier rules by hand and tuning them manually, their automatic optimization will lead to improved results for complex control
strategies.

9.7   Summary

Numerous applications have proven the success of GENETIC ALGORITHMS in practical applications. An important aspect for this success story is their broad applicability.
Once arrived in the world of GENETIC ALGORITHMS, it sounds appealing to model
every optimization problem with fitness functions and to adapt and employ GENETIC
ALGORITHMS for solving them. Also the design of appropriate genetic operators and
the choice of parameters is usually a convenient task. Moreover, the most successful
GENETIC ALGORITHMS in applications incorporate expert knowledge. This includes
initialization procedures, for example the use of solutions that are already known
as parents in the first generation, and the modification of solution parts with expert
designs. In applications the practitioner can be part of an interactive optimization
loop. The visualization and even sonification of GENETIC ALGORITHM runs can
support the integration of human decisions into an automated optimization process.

In this chapter we have shown successful examples for the application of GENETIC
ALGORITHMS in dimensionality reduction, classifier ensembles, feature tuning for
wind power prediction, wind turbine placement, and learning classifier systems. It
will be interesting to see the number of successful applications growing in the future.

Part IV
Ending

Chapter 10
Summary and Outlook

10.1 Summary

GENETIC ALGORITHMS belong to the most important and successful algorithms in learning and optimization. For difficult optimization problems GENETIC ALGORITHMS are excellent solution strategies. Based on the concept of iteratively approximating a solution, they are applicable to a broad class of problems. Their relation to natural evolution make them interesting and powerful algorithms. For beginners the evolutionary concepts are attractive for learning and understanding of optimization processes. For researchers borrowing novel concepts from evolution is an inexhaustible source of inspiration. GENETIC ALGORITHMS are typically used in an online setting when optimizing an unknown problem. This means that the practitioner is part of the search process by trying different settings, parameters, and algorithmic variants.

Numerous extensions have been proposed in the past. We concentrated on a selection of the most important algorithms and concepts for GENETIC ALGORITHMS to search in multimodal, constrained, and multi-objective solution spaces. Extensions for multimodal solution spaces are based on niches and diversity maintenance. Constraint handling techniques allow dealing with restrictions that have to be considered during optimization. Multi-objective techniques allow the approximation of a whole set of Pareto-optimal solutions at once.

Meanwhile, a remarkable set of theoretical tools and analyses for GENETIC ALGORITHMS has been proposed. GENETIC ALGORITHMS might have started from a weakly understood cradle of heuristic methods, but have meanwhile grown to a rich set of methods with theoretical and practical support. Further, GENETIC ALGORITHMS profit from other branches of computer science and artificial intelligence like machine learning. If the search is equipped with methods from learning, for example the concept of fitness and constraint surrogates, the optimization process is often significantly improved. Covariance matrix estimation supports the search in continuous solution spaces while dimensionality reduction methods can be employed for visualization of high-dimensional optimization processes.

O. Kramer, *Genetic Algorithm Essentials*, Studies in Computational
Intelligence 679, DOI 10.1007/978-3-319-52156-5_10

10.2 Outlook

GENETIC ALGORITHMS will surely play an important role in many optimization settings in the future. The broad set of methods that have been developed for a huge variety of problem classes will be applicable to an enormous set of applications in the future. Moreover, GENETIC ALGORITHMS will continue their development. Besides the extension of GENETIC ALGORITHMS with heuristic mechanisms, which adapt them to specific problem classes, various applications will play a significant role for their growth. In Chap. 9 we already presented a small set of potential application domains. Hundreds and even thousands more exist in literature and will also be subject to future research and developments.

A promising yet surprisingly old research area is parallelization. GENETIC ALGORITHMS are easy to parallelize, in particular, if numerous solutions can be generated and evaluated at the same time. Parallelization on graphical processing units (GPUs) celebrates its success with deep learning currently, but also approaches for GENETIC ALGORITHMS have successfully been introduced [34]. Deep learning is a revolutionary technique as an extension of artificial neural networks to architectures with many layers and hundreds of neurons. We expect a similar success for GENETIC ALGORITHMS in the near future based on massive parallelization, also in the area of evolving deep learning networks with GENETIC ALGORITHMS.

We hope that this book has served as a convenient introduction and overview to GENETIC ALGORITHMS and will thus contribute to a fruitful development in GENETIC ALGORITHM-based research in the future. The depiction will continuously be improved and updated to capture latest developments in research and applications.

References

1. T. Bäck and M. Schüctz. Intelligent mutation rate control in canonical genetic algorithms. In *Foundation of Intelligent Systems*, pages 158–167. Springer, 1996.
2. S. H. Bak, N. Rask, and S. Risi. Towards adaptive evolutionary architecture. In *Evolutionary and Biologically Inspired Music, Sound, Art and Design (EvoStar)*, pages 47–62, 2016.
3. W. Banzhaf, F. D. Francone, R. E. Keller, and P. Nordin. *Genetic Programming: An Introduction: on the Automatic Evolution of Computer Programs and Its Applications*. Morgan Kaufmann Publishers Inc., San Francisco, CA, USA, 1998.
4. N. Beume, B. Naujoks, and M. Emmerich. SMS-EMOA: Multiobjective Selection based on Dominated Hypervolume. *European Journal of Operational Research*, 181(3):1653–1669, 2007.
5. H. Beyer and H. Schwefel. Evolution strategies - A comprehensive introduction. *Natural Computing*, 1(1):3–52, 2002.
6. H. G. Beyer. On the performance of $(1, \lambda)$-evolution strategies for the ridge function class. *Transactions on Evolutionary Computation*, 5(3):218–235, 2001.
7. E. Bonabeau, M. Dorigo, and G. Theraulaz. *Swarm Intelligence: From Natural to Artificial Systems*. Oxford University Press, Inc., New York, NY, USA, 1999.
8. B. E. Boser, I. M. Guyon, and V. N. Vapnik. A training algorithm for optimal margin classifiers. In *Workshop on Computational Learning Theory (COLT)*, pages 144–152, New York, NY, USA, 1992. ACM.
9. L. Breiman. Random forests. *Machine Learning*, 45(1):5–32, 2001.
10. J. Bremer and S. Lehnhoff. A decentralized PSO with decoder for scheduling distributed electricity generation. In *Applications of Evolutionary Computation (EvoStar)*, pages 427–442, 2016.
11. J. Chen, Q. Yang, J. Ni, Y. Xie, and S. Cheng. An improved fireworks algorithm with landscape information for balancing exploration and exploitation. In *IEEE Congress on Evolutionary Computation (CEC)*, pages 1272–1279, 2015.
12. R. Clausen, E. Sapin, K. A. D. Jong, and A. Shehu. Evolution strategies for exploring protein energy landscapes. In *Genetic and Evolutionary Computation Conference (GECCO)*, pages 217–224, 2015.
13. C. Darwin. *On the Origin of Species*. John Murray, London, 1859.
14. E. Davies, P. Tew, D. R. Glowacki, J. Smith, and T. Mitchell. Evolving atomic aesthetics and dynamics. In *Evolutionary and Biologically Inspired Music, Sound, Art and Design (EvoStar)*, pages 17–30, 2016.
15. W. de Landgraaf, A. Eiben, and V. Nannen. Parameter calibration using meta-algorithms. In *IEEE Congress on Evolutionary Computation (CEC)*, pages 71–78, 2007.

© Springer International Publishing AG 2017

O. Kramer, *Genetic Algorithm Essentials*, Studies in Computational Intelligence 679, DOI 10.1007/978-3-319-52156-5

16. W. A. de Landgraaf, A. E. Eiben, and V. Nannen. Parameter calibration using meta-algorithms. In *IEEE Congress on Evolutionary Computation (CEC)*, pages 71–78, 2007.

17. K. Deb, S. Agrawal, A. Pratap, and T. Meyarivan. A fast and elitist multiobjective genetic algorithm: NSGA-II. *IEEE Transactions on Evolutionary Computation*, 6(2):182–197, 2002.

18. B. Desjardins, R. Falcon, R. S. Abielmona, and E. M. Petriu. A multi-objective optimization approach to reliable robot-assisted sensor relocation. In *IEEE Congress on Evolutionary Computation (CEC)*, pages 956–964, 2015.

19. B. Doerr and C. Doerr. A tight runtime analysis of the $(1+(\lambda, \lambda))$ genetic algorithm on onemax. In *Genetic and Evolutionary Computation Conference (GECCO)*, pages 1423–1430, 2015.

20. M. Dorigo, V. Maniezzo, and A. Colorni. Ant system: Optimization by a colony of cooperating agents. *Transactions on Systems, Man, and Cybernetics*, 26(1):29–41, 1996.

21. A. E. Eiben, E. H. L. Aarts, and K. M. van Hee. Global convergence of genetic algorithms: A markov chain analysis. In *Parallel Problem Solving from Nature (PPSN)*, pages 4–12, Berlin, 1991. Springer.

22. A. E. Eiben, R. Hinterding, and Z. Michalewicz. Parameter control in evolutionary algorithms. *Transactions on Evolutionary Computation*, 3(2):124–141, 1999.

23. A. E. Eiben and J. E. Smith. *Introduction to Evolutionary Computing*. Natural Computing Series. Springer, Berlin, 2003.

24. A. Emami and P. Noghreh. New approach on optimization in placement of wind turbines within wind farm by genetic algorithms. *Renewable Energy*, 35(7):1559–1564, 2010.

25. M. Ester, H.-P. Kriegel, J. Sander, and X. Xu. A density-based algorithm for discovering clusters in large spatial databases with noise. In *International Conference on Knowledge Discovery and Data Mining (KDD)*, pages 226–231. AAAI Press, 1996.

26. T. C. Fogarty. Varying the probability of mutation in the genetic algorithm. In *International Conference on Genetic Algorithms*, pages 104–109, San Francisco, 1989. Morgan Kaufmann Publishers Inc.

27. L. Fogel, A. A.J. Owens, and M. Walsh. *Artificial Intelligence through Simulated Evolution*. Wiley, 1971.

28. S. Forrest and M. Mitchell. Relative building-block fitness and the building block hypothesis. In *Foundations of Genetic Algorithms (FOGA)*, pages 109–126, 1992.

29. T. Friedrich, T. Kroeger, and F. Neumann. Weighted preferences in evolutionary multi-objective optimization. *International Journal of Machine Learning and Cybernetics*, 4(2):139–148, 2013.

30. C. Gießen and C. Witt. Population size vs. mutation strength for the $(1+\lambda)$ EA on onemax. In *Genetic and Evolutionary Computation Conference (GECCO)*, pages 1439–1446, 2015.

31. D. E. Goldberg. *Genetic Algorithms in Search, Optimization, and Machine Learning*. Addison Wesley, 1989.

32. D. E. Goldberg and R. Lingle. Alleles, loci, and the traveling salesman problem. In *International Conference on Genetic Algorithms and Their Applications*, pages 154–159, 1985.

33. J. S. Gonzalez, A. Gonzalez Rodriguez, J. C. Mora, J. R. Santos, and M. B. Payan. Optimization of wind farms using an evolutive algorithm. *Renewable Energy*, 35(8):1671–1681, 2010.

34. S. Gupta and G. Tan. A scalable parallel implementation of evolutionary algorithms for multi-objective optimization on gpus. In *IEEE Congress on Evolutionary Computation (CEC)*, pages 1567–1574, 2015.

35. M. M. Haklay and P. Weber. Openstreetmap: User-generated street maps. *IEEE Pervasive Computing*, 7(4):12–18, 2008.

36. N. Hansen and A. Ostermeier. Adapting arbitrary normal mutation distributions in evolution strategies: The covariance matrix adaptation. In *IEEE Congress on Evolutionary Computation (CEC)*, pages 312–317, 1996.

37. N. Hansen and A. Ostermeier. Completely derandomized self-adaptation in evolution strategies. *Evolutionary Computation*, 9(2):159–195, 2001.

38. T. Hastie, R. Tibshirani, and J. Friedman. *The Elements of Statistical Learning*. Springer Series in Statistics. Springer, New York, Heidelberg, 2009.

39. J. Holland. *Adaptation in Natural and Artificial Systems*. MIT Press, 1975.

40. J. Holland. *Hidden Order: How Adaptation Builds Complexity.* Helix Books, 1996.
41. J. D. Hunter. Matplotlib: A 2d graphics environment. *Computing In Science & Engineering,* 9(3):90–95, 2007.
42. G. James, D. Witten, T. Hastie, and R. Tibshirani. *An Introduction to Statistical Learning.* Springer Texts in Statistics. Springer, New York, Heidelberg, 2013.
43. T. Jansen and I. Wegener. Real royal road functions-where crossover provably is essential. *Discrete Applied Mathematics,* 149(1–3):111–125, 2005.
44. J. Joines and C. Houck. On the use of non-stationary penalty functions to solve nonlinear constrained optimization problems with GAs. In D. B. Fogel, editor, *IEEE Congress on Evolutionary Computation (CEC),* pages 579–584, Orlando, Florida, 1994. IEEE Press.
45. I. T. Jolliffe. *Principal component analysis.* Springer Series in Statistics. Springer, New York u.a., 1986.
46. G. Karafotias, M. Hoogendoorn, and Á. E. Eiben. Parameter control in evolutionary algorithms: Trends and challenges. *Transactions on Evolutionary Computation,* 19(2):167–187, 2015.
47. J. Kennedy and R. Eberhart. Particle swarm optimization. In *IEEE International Conference on Neural Networks (IJCNN),* pages 1942–1948, 1995.
48. S. Kirkpatrick, C. G. Jr, and M. Vecchi. Optimization by simulated annealing. *Science,* 220(4598):671–680, 1983.
49. J. R. Koza. *Genetic Programming: On the Programming of Computers by Means of Natural Selection.* MIT Press, Cambridge, 1992.
50. O. Kramer. *Self-Adaptive Heuristics for Evolutionary Computation,* volume 147 of *Studies in Computational Intelligence.* Springer, 2008.
51. O. Kramer. Evolutionary self-adaptation: a survey of operators and strategy parameters. *Evolutionary Intelligence,* 3(2):51–65, 2010.
52. O. Kramer. Iterated local search with powell's method: a memetic algorithm for continuous global optimization. *Memetic Computing,* 2(1):69–83, 2010.
53. O. Kramer. *Dimensionality Reduction with Unsupervised Nearest Neighbors,* volume 51 of *Intelligent Systems Reference Library.* Springer, 2013.
54. O. Kramer. Evolution strategies with ledoit-wolf covariance matrix estimation. In *IEEE Congress on Evolutionary Computation (CEC),* pages 1712–1716, 2015.
55. O. Kramer. Local fitness meta-models with nearest neighbor regression. In *Applications of Evolutionary Computation (EvoStar),* pages 3–10, 2016.
56. O. Kramer. *Machine Learning for Evolution Strategies,* volume 20 of *Studies in Big Data.* Springer, 2016.
57. O. Kramer and H. Danielsiek. Dbscan-based multi-objective niching to approximate equivalent pareto-subsets. In *Genetic and Evolutionary Computation Conference (GECCO),* pages 503–510, 2010.
58. O. Kramer and P. Koch. Rake selection: A novel evolutionary multi-objective optimization algorithm. In *Advances in Artificial Intelligence (KI),* pages 177–184, 2009.
59. O. Kramer and D. Lückehe. Visualization of evolutionary runs with isometric mapping. In *IEEE Congress on Evolutionary Computation (CEC),* pages 1359–1363, 2015.
60. O. Kramer, B. Satzger, and J. Lässig. Managing energy in a virtual power plant using learning classifier systems. In *International Conference on Genetic and Evolutionary (GEM),* pages 111–117, 2010.
61. O. Kramer, U. Schlachter, and V. Spreckels. An adaptive penalty function with meta-modeling for constrained problems. In *IEEE Congress on Evolutionary Computation (CEC),* pages 1350–1354, 2013.
62. O. Kramer and H. Schwefel. On three new approaches to handle constraints within evolution strategies. *Natural Computing,* 5(4):363–385, 2006.
63. J. W. Kruisselbrink, E. Reehuis, A. H. Deutz, T. Bäck, and M. Emmerich. Using the uncertainty handling CMA-ES for finding robust optima. In *Genetic and Evolutionary Computation Conference (GECCO),* pages 877–884, 2011.

64. A. Kusiak and Z. Song. Design of wind farm layout for maximum wind energy capture. *Renewable Energy*, 35(3):685–694, 2010.
65. P. Larranaga and J. Lozano. *Estimation of Distribution Algorithms. A New Tool for Evolutionary Computation*. Kluwer Academic Publishers, 2001.
66. J. A. Lee and M. Verleysen. *Nonlinear Dimensionality Reduction*. Springer Series in Statistics. Springer, 2007.
67. D. Lessin and S. Risi. Darwin's avatars: A novel combination of gameplay and procedural content generation. In *Genetic and Evolutionary Computation Conference (GECCO)*, pages 329–336, 2015.
68. X. Li, S. Zeng, S. Qin, and K. Liu. Constrained optimization problem solved by dynamic constrained NSGA-III multiobjective optimizational techniques. In *IEEE Congress on Evolutionary Computation (CEC)*, pages 2923–2928, 2015.
69. D. P. Liebana, J. Dieskau, M. Hunermund, S. Mostaghim, and S. M. Lucas. Open loop search for general video game playing. In *Genetic and Evolutionary Computation Conference (GECCO)*, pages 337–344, 2015.
70. I. Loshchilov. CMA-ES with restarts for solving CEC 2013 benchmark problems. In *IEEE Congress on Evolutionary Computation (CEC)*, pages 369–376, 2013.
71. I. Loshchilov. A computationally efficient limited memory CMA-ES for large scale optimization. In *Genetic and Evolutionary Computation Conference (GECCO)*, pages 397–404, 2014.
72. I. Loshchilov, M. Schoenauer, and M. Sebag. Self-adaptive surrogate-assisted covariance matrix adaptation evolution strategy. In *Genetic and Evolutionary Computation Conference (GECCO)*, pages 321–328, 2012.
73. D. Lückehe and O. Kramer. Leaving local optima in unsupervised kernel regression. In *Artificial Neural Networks and Machine Learning (ICANN)*, pages 137–144, 2014.
74. D. Lückehe, M. Wagner, and O. Kramer. On evolutionary approaches to wind turbine placement with geo-constraints. In *Genetic and Evolutionary Computation Conference (GECCO)*, pages 1223–1230, 2015.
75. D. Lückehe, M. Wagner, and O. Kramer. Constrained evolutionary wind turbine placement with penalty functions. In *IEEE Congress on Evolutionary Computation (CEC)*, pages 4903–4910, 2016.
76. P. Meinicke, S. Klanke, R. Memisevic, and H. Ritter. Principal surfaces from unsupervised kernel regression. *Transactions on Pattern Analysis and Machine Intelligence*, 27(9):1379–1391, 2005.
77. I. Mierswa. Controlling overfitting with multi-objective support vector machines. In *Genetic and Evolutionary Computation Conference (GECCO)*, pages 1830–1837, 2007.
78. Á. Monteagudo and J. S. Reyes. Evolutionary optimization of cancer treatments in a cancer stem cell context. In *Genetic and Evolutionary Computation Conference (GECCO)*, pages 233–240, 2015.
79. G. Morse and K. O. Stanley. Simple evolutionary optimization can rival stochastic gradient descent in neural networks. In *Genetic and Evolutionary Computation Conference (GECCO)*, pages 477–484, 2016.
80. F. Neumann and C. Witt. *Bioinspired Computation in Combinatorial Optimization: Algorithms and Their Computational Complexity*. Natural Computing Series. Springer, Berlin, 2010.
81. S. Oehmcke, J. Heinermann, and O. Kramer. Analysis of diversity methods for evolutionary multi-objective ensemble classifiers. In *Applications of Evolutionary Computation (EvoStar)*, pages 567–578, 2015.
82. A. Ostermeier, A. Gawelczyk, and N. Hansen. Step-size adaption based on non-local use of selection information. In *Parallel Problem Solving from Nature (PPSN)*, pages 189–198, 1994.
83. K. Pearson. On lines and planes of closest fit to systems of points in space. *Philosophical Magazine*, 2(6):559–572, 1901.

84. F. Pedregosa, G. Varoquaux, A. Gramfort, V. Michel, B. Thirion, O. Grisel, M. Blondel, P. Prettenhofer, R. Weiss, V. Dubourg, J. Vanderplas, A. Passos, D. Cournapeau, M. Brucher, M. Perrot, and E. Duchesnay. Scikit-learn: Machine learning in Python. *Journal of Machine Learning Research*, 12:2825–2830, 2011.

85. S. Peignier, C. Rigotti, and G. Beslon. Subspace clustering using evolvable genome structure. In *Genetic and Evolutionary Computation Conference (GECCO)*, pages 575–582, 2015.

86. I. Rechenberg. *Evolutionsstrategie - Optimierung technischer Systeme nach Prinzipien der biologischen Evolution*. Fromman-Holzboog, 1971.

87. F. H. Rios, L. König, and H. Schmeck. Stigmergy-based scheduling of flexible loads. In *Applications of Evolutionary Computation (EvoStar)*, pages 475–490, 2016.

88. F. Rosenblatt. The perceptron. a probabilistic model for information storage and organization in the brain. *Psychological Reviews*, 65:386–408, 1958.

89. H. Rosenbrock. An automatic method for finding the greatest or least value of a function. *The Computer Journal*, 3(3):175–184, 1960.

90. S. T. Roweis and L. K. Saul. Nonlinear dimensionality reduction by locally linear embedding. *Science*, 290:2323–2326, 2000.

91. G. Rudolph. Finite markov chain results in evolutionary computation: A tour d'horizon. *Fundamenta Informaticae*, 35(1-4):67–89, 1998.

92. B. Schölkopf and A. J. Smola. *Learning with Kernels*. MIT Press, 2002.

93. H. Schwefel. *Numerische Optimierung von Computer-Modellen*. Birkhäuser, 1977.

94. H. Schwefel. *Evolution and Optimum Seeking*. Wiley, New York, 1995.

95. M. Scirea, J. Togelius, P. W. Eklund, and S. Risi. Metacompose: A compositional evolutionary music composer. In *Evolutionary and Biologically Inspired Music, Sound, Art and Design (EvoStar)*, pages 202–217, 2016.

96. H. Seada and K. Deb. Effect of selection operator on NSGA-III in single, multi, and many-objective optimization. In *IEEE Congress on Evolutionary Computation (CEC)*, pages 2915–2922, 2015.

97. G. Syswerda. Simulated crossover in genetic algorithms. In *Foundations of Genetic Algorithms (FOGA)*, pages 239–255, 1992.

98. Y. Tan and Y. Zhu. Fireworks algorithm for optimization. In *Advances in Swarm Intelligence*, pages 355–364, 2010.

99. M. A. M. Teixeira, F. Goulart, and F. Campelo. Evolutionary multiobjective optimization of winglets. In *Genetic and Evolutionary Computation Conference (GECCO)*, pages 1021–1028, 2016.

100. J. B. Tenenbaum, V. D. Silva, and J. C. Langford. A global geometric framework for nonlinear dimensionality reduction. *Science*, 290:2319–2323, 2000.

101. N. A. Treiber and O. Kramer. Evolutionary feature weighting for wind power prediction with nearest neighbor regression. In *IEEE Congress on Evolutionary Computation (CEC)*, pages 332–337, 2015.

102. N. A. Treiber, S. Späth, J. Heinermann, L. von Bremen, and O. Kramer. Comparison of numerical models and statistical learning for wind speed prediction. In *European Symposium on Artificial Neural Networks (ESANN)*, pages 71–76, 2015.

103. V. N. Vapnik. *The Nature of Statistical Learning Theory*. Springer, New York, 1995.

104. F. Wilcoxon. Individual comparisons by ranking methods. *Biometrics Bulletin*, 1:80–83, 1945.

105. D. H. Wolpert and W. G. Macready. No free lunch theorems for search. Technical report, Santa Fe, 1995.

106. X.-S. Yang. *Nature-Inspired Metaheuristic Algorithms*. Luniver Press, 2008.

107. E. Zitzler and L. Thiele. Multiobjective Evolutionary Algorithms: A Comparative Case Study and the Strength Pareto Approach. *Transactions on Evolutionary Computation*, 3(4):257–271, 1999.

Index

© Springer International Publishing AG 2017
O. Kramer, *Genetic Algorithm Essentials*, Studies in Computational
Intelligence 679, DOI 10.1007/978-3-319-52156-5

Printed in the United States
By Bookmasters